Optical Technologies for Advancing Communication, Sensing, and Computing Systems

Edited by Sulaiman Wadi Harun

Published in London, United Kingdom

Optical Technologies for Advancing Communication, Sensing, and Computing Systems
http://dx.doi.org/10.5772/intechopen.1003549
Edited by Sulaiman Wadi Harun

Contributors
Ahmed Hisham E. Morshed, Alexander S. Sigov, Bipin Kumar Saw, Conrad Bertrand Tabi, Eugeniy Plastinin, Gaston Camus Latchio Tiofack, Ghanshyam Singh, Hippolyte Tagwo, Karabo Kefilwe Ndebele, Manish Tiwari, Mikhail E. Belkin, Mikhail G. Vasil'ev, Muhammad Nayab Ahmad, Noama Ouchani, Olga N. Egorova, Ricardo Rovere de Santi, Sergei L. Semjonov, Sujit K. Bose, Timoléon Crépin Kofane, Uzma Hira

First published in London, United Kingdom, 2025 by IntechOpen
IntechOpen is the global imprint of INTECHOPEN LIMITED, registered in England and Wales, registration number: 11086078, 167-169 Great Portland Street, London, W1W 5PF, United Kingdom

For EU product safety concerns: IN TECH d.o.o., Prolaz Marije Krucifikse Kozulić 3, 51000 Rijeka, Croatia, info@intechopen.com or visit our website at intechopen.com.

British Library Cataloguing-in-Publication Data
A catalogue record for this book is available from the British Library

Optical Technologies for Advancing Communication, Sensing, and Computing Systems
Edited by Sulaiman Wadi Harun
p. cm.
Print ISBN 978-0-85466-155-8
Online ISBN 978-0-85466-154-1
eBook (PDF) ISBN 978-0-85466-156-5

Meet the editor

Sulaiman Wadi Harun received his B.E. degree in Electrical and Electronics System Engineering from Nagaoka University of Technology, Japan, in 1996 and his M.Sc. and Ph.D. degrees in Photonics Technology from the University of Malaya in 2001 and 2004, respectively. With over 20 years of research experience, he has contributed extensively to developing optical fiber devices, including fiber amplifiers, fiber lasers, and fiber optic sensors. He is also exploring new nanomaterials such as graphene, carbon nanotubes, black phosphorus, topological insulators, and zinc oxide nanostructures for fiber lasers and sensor applications. Prof. Harun has authored over 900 articles in renowned ISI journals, with his work cited more than 9,000 times and an h-index exceeding 55, reflecting his significant impact on the field. He is an academic editor for the International Journal of Optics, indexed in ISI and Scopus. He has successfully supervised more than 80 Ph.D. students to completion. Additionally, he holds adjunct professorships at Brawijaya University in Indonesia and Vellore Institute of Technology in India. In 2016, he was awarded the prestigious Malaysian Rising Star award by the Ministry of Higher Education for his contributions to international collaboration. He is a Fellow of the Malaysian Academy of Science and the founder and honorary advisor of the Optical Society of Malaysia.

Contents

Preface

An optical fiber used for data transmission typically consists of a core and cladding, with the cladding having a slightly lower refractive index than the core to enable light propagation through total internal reflection. The core can either be optically homogeneous or modified through advanced processing to enhance transmission efficiency. Chapter 1 presents the theory of light-wave dispersion and attenuation as it propagates along the fiber. The development of this theory is based on Maxwell's equations, with the solution represented by the Hertz vector Π, oriented along the fiber core axis. For a homogeneous core step-index fiber, Π satisfies the wave equation for light propagation. At the same time, a slight modification of the representation yields a modified wave equation for a graded-index inhomogeneous core. The chapter also investigates the effects of macro-bending for small bends, finding that such bends result in power loss and thus should be minimized during cable installation. Chapter 2 explores the effects of higher-order Raman scattering on modulational instability in optical fibers with pure-quartic dispersion. Self-frequency shift effects influence both low and high input power regimes. Due to the presence of quintic nonlinearity and quartic dispersion, this results in extended regions of instability within the gain spectrum. Numerical simulations demonstrate that the biharmonic nonlinear Schrödinger equation generates ultrashort pulse trains, including rogue wave-like patterns and Akhmediev breather trains. This finding paves the way for a deeper understanding of the impact of higher-order Raman scattering terms on the generation and propagation of pure-quartic solitons, with potential applications in silica photonic crystals and photonic crystal waveguides.

Optical fiber sensors have become crucial technological advancement due to their exceptional sensitivity, resistance to electromagnetic interference, and durability in challenging environments. They are widely used across industries such as environmental sensing, structural health monitoring, and medical diagnostics. Chapter 3 emphasizes the transformative impact of optical fiber sensors by exploring their technological developments, potential innovations, and prospects. This chapter covers various types of optical fiber sensors, including fiber Bragg grating interferometers and biosensors. Intensity-based fiber sensors are of significant interest in both industrial and research settings due to their simplicity and cost-effectiveness. A subset of these sensors uses wave interference in fiber structures to induce intensity variations, which are then detected. These in-fiber interferometers leverage phase-sensitive wave variations to achieve higher sensitivity than pure intensity-based sensors. Chapter 4 explores optical fiber sensors based on modal interference in multimode optical fibers, covering their operating principles, potential applications, and challenges for industrial implementation. It reviews and contrasts various sensor structures and sensing approaches and discusses sensor applications and multiplexing schemes.

Chapter 5 presents an analytical review of scientific literature supporting the feasibility, benefits, and effectiveness of incorporating spatial multiplexing technology in fiber-optic communication systems. This technology is particularly relevant for rapidly developing sectors such as 5th-generation cellular networks in megacities and interconnecting links for hyper-scale data centers. The chapter explores optimal design principles and outlines the key requirements for transmission and reception equipment components and units. It describes the development of a conceptual schematic for interconnecting communication links between hyper-scale data centers, utilizing time, spectral, and spatial multiplexing with multi-core optical fibers and connectors. The chapter also highlights the achievement of a record throughput of over 4 Tbit/s and specific energy consumption of no more than 0.1 nJ/bit. The chapter discusses the results from computer modeling, prototyping, and experimental studies on a fiber-optic path based on a 7-core optical fiber to validate the proposed design.

Quantum computing has long promised to revolutionize computational power and cryptographic security. Still, progress has been slow due to the immense challenges in developing these systems. One major hurdle was the need for operation at near-absolute zero temperatures, which restricted practical applications and limited quantum processors to only a thousand qubits. Chapter 6 briefly reviews recent advancements in using optical fibers to improve readout capabilities and increase the number of physical qubits in superconducting quantum computers.

Free Space Optical (FSO) communication, which uses light to transmit information, has gained significant attention for its potential in high-speed, secure, and license-free communication. Chapter 7 reviews various modulation schemes used in FSO systems, including Phase Shift Keying (PSK), Orthogonal Frequency Division Multiplexing (OFDM), Space Shift Keying (SSK), Pulse Position Modulation (PPM), Gaussian Minimum Shift Keying (GMSK), and Subcarrier Intensity Modulation (SIM). These schemes modify different aspects of the signal: PSK alters phase, OFDM introduces orthogonal subcarriers, SSK shifts beam location, PPM changes pulse positions, GMSK applies a Gaussian filter to phase shifts, and SIM adjusts subcarrier amplitudes. Each modulation scheme is optimized for specific transmission conditions, balancing data rate and reliability depending on factors like system complexity, bandwidth availability, and atmospheric conditions.

Chapter 8 theoretically explores the potential of utilizing 1D birefringent dielectric photonic crystals in the design of optical devices. It demonstrates that a 1D structure can be used to create a transverse-electric/transverse-magnetic polarization converter. The study also reveals that, with the right choice of material and geometric properties, a birefringent photonic crystal can exhibit an absolute photonic band gap, where light propagation is prohibited for all polarizations and a given plane of incidence. Additionally, the chapter examines the use of a single-defect anisotropic photonic crystal for fabricating an optical switch and a selective electromagnetic filter. The birefringence of the photonic crystal layers enhances the performance of these optical devices compared to isotropic structures.

Sulaiman Wadi Harun
Department of Electrical Engineering,
University of Malaya,
Kuala Lumpur, Malaysia

Chapter 1

Theory of Dispersion and Attenuation of Light Wave Propagation in Fiber-Optic Cables

Sujit K. Bose

Abstract

A fiber-optic cable consists of one or more optical fibers having slightly less refractive index for guiding the light wave. The central core of a fiber is either optically homogeneous or rendered inhomogeneous by technical processing for greater efficiency in transmission. The theory of dispersion of light waves through such types of fibers is presented here, based on the Maxwell equations by representing the EM field by the single Hertz vector $\mathbf{\Pi}$ instead of the electric and magnetic intensities $\mathbf{E}$ and $\mathbf{H}$. The study of the phenomenon is presented in three parts. At first, a straight step-index fiber with homogeneous core and cladding is considered, showing considerable dispersion of propagating waves leading to formation of wave groups that might even burst. The effect of small bending of a cable line is analyzed next, and it is shown that attenuation proportional to the magnitude of curvature takes place in the propagation process. Finally, inhomogeneity of the fiber core is treated, obtaining a modified wave equation for $\mathbf{\Pi}$, solving it for the parabolic index core of some degree α. Numerical study for the case $\alpha = 2$, showed marked reduction in dispersion. The method is then generalised to a general index core and applied to a non-smooth profile of NZDSF type showing propagation without dispersion.

Keywords: fiber-optic cable, step-index and graded-index cores, dispersion effects, macrobending attenuation, hertz vector

1. Introduction

The transmission of packets of data in optical fibers takes place as modulated pulses of monochromatic light waves guided through the core of the fiber, having a thin cladding of slightly less refractive index than that of the core material, to ensure total internal refraction of the passing light wave. A jacket over the composite media provides protection from physical damage to the optical carrier. In practice, a cable may contain more than one such fiber for enhanced technical performance. The medium of the core of a fiber is generally optically homogeneous, but increasingly especially designed inhomogeneous cores are being developed for increasing the efficiency of transmission. The two types are generally known as step-index and graded-index optical fibers. Theoretical study of both the types is presented in most books on fiber optics [1–5]. The methodology of treatment is twofold: A preliminary geometrical optics approach followed by detailed wave guide action through the core based on Maxwell equations of electromagnetism. The validity of the latter approach is well explained in the Introduction of Born and Wolf ([6], p. xxx), stating that light waves are electromagnetic (EM) waves in the frequency range of 175–375 THz with wavelength between 770 and 1675 nm. As a consequence, most writings on the subject employ the electric intensity E and the magnetic intensity H for the EM field.

In the exposition presented here, the wave guide action through the core is treated by using a single vector field Π—the Hertz vector, followed by systematic analysis of three specific problems. At first, a straight step-index optical fiber is treated adopting that formalism, obtaining in a much simpler manner the well-known dispersion equation mentioned in the References [1–5]. The same equation was derived by Bose [7] following a more elaborate solution given in Stratton [8]. In that paper, it was shown that vanishing of the group velocity of the dispersed wave can cause catastrophic "bursts" for certain frequencies and phase velocities. Such abrupt behavior can occur in reality, as it has been observed in the high-resolution experiments of Willinger et al. [9], Thompson et al. [10], and Kim and Won [11]. Next, the problem of bends in the fibers is studied, as bends are inevitable in the laying of long fiber-optic cables. For the purpose of analysis, the bends are assumed to be gradually varying in curvature. By an approximate analysis, it is found that the wave number k of propagation has a negative imaginary part proportional to the magnitude of the curvature, with the implication that there is a linear "macrobending loss" of power of the light beam in the core. This is in complete agreement with the finding in Bose [12] for a circularly bent cable, when the radius of the bend exceeds a certain critical value. In the final section, the fundamental equations for a straight graded-index fiber are developed, again in terms of the Hertz vector Π defined slightly differently for the purpose. This section closely follows the method presented by Bose [13] for an analysis of the problem and applied to a core with a parabolic profile of general degree α. Computations are carried out for the dispersion equation for the case $\alpha = 2$, and it is found that the propagation has reduced dispersion to quite an extent. This formalism is generalized to any form of the variable refractive index of the core material in the next section, treating the special case of a profile similar to that of the Nonzero Dispersion Shifted Fiber (NZDSF) as a numerical case. The importance of optics and optical communication in recent times can be gauged by the appearance of recent comprehensive articles by Miller [14] on optics and by Cao et al. [15] on multimode optical fibers, containing numerous other references.

2. Propagation in a step-index fiber

A step-index optical fiber consists of a homogeneous core all along its length with a certain constant refractive index n_1, with a very thin doped cladding of some refractive index n_2 slightly less than that of the core, viz. n_1. A modulated pulse of digitally coded information light wave, propagating with some nearly fixed frequency, guided through the core, can in general be decomposed into a number of modes by Fourier's theorem ([6], p. 10) depending on the diameter of the core. Even though multimodal cables are deployed in practice, unimodal transmission consisting of only the fundamental least frequency mode is often employed due to some technical advantages.

Light wave propagation is governed by the Maxwell equations in terms of an electric intensity E and a magnetic intensity H. An account of the modal analysis of the associated electromagnetic waves in a circular cylinder of radius a embedded in an infinite homogeneous medium is given in Stratton ([8], p. 524). Such a model of the wave guide action for a fiber-optic cable fits the portrayal, because of the fact that the propagation is confined to the core region due to total internal reflection. The theoretical development of the modal propagation becomes quite simple on account of adopting such a model. Bose [7] presented an analysis of the problem based on the very general treatment of Stratton [8], to show that the propagating wave shows a dispersive character leading to possible adverse effects. In the following, a much simpler treatment is presented for the propagation characteristics.

The Maxwell equations in the absence of any electric charge and current in the medium are:

$$\nabla \cdot \mathbf{D} = 0 \tag{1}$$

$$\nabla \cdot \mathbf{B} = 0 \tag{2}$$

$$\nabla \times \mathbf{E} = -\frac{\partial \mathbf{B}}{\partial t} \tag{3}$$

$$\nabla \times \mathbf{H} = \frac{\partial \mathbf{D}}{\partial t} \tag{4}$$

where the displacement current D and the magnetic induction B are given by the equations

$$\mathbf{D} = \epsilon \mathbf{E} \tag{5}$$

and

$$\mathbf{B} = \mu \mathbf{H} \tag{6}$$

ϵ, μ being the electric permittivity and the magnetic permeability such that $\epsilon\mu = 1/c^2$, c = velocity of light in the medium, depending upon its refractive index n. In the core medium with index n_1, $c_1 = c_0/n_1$ where c_0 = velocity of light in vacuum. Similarly in the cladding material with index n_2, $c_2 = c_0/n_2$. Since $n_1 > n_2$, it follows that $c_1 < c_2$. The electromagnetic vector field (E, H) is conveniently represented by a single vector field Π—the Hertz vector defined as follows ([8], p. 28). Since by Eq. (2) the induction vector B is solenoidal, one can write

$$\mathbf{B} = \mu\epsilon\nabla \times \dot{\Pi} \quad \text{(curl of vector field)} \tag{7}$$

where the dot denotes differentiation with respect to time t. Thus, from Eq. (6) it follows that

$$\mathbf{H} = \epsilon\nabla \times \dot{\Pi} \tag{8}$$

Inserting the expression for H given by Eq. (8) in Eq. (4), using Eq. (5), it follows that

$$\mathbf{E} = \nabla \times \nabla \times \Pi \tag{9}$$

Now inserting the above expression for E in Eq. (4), using Eqs. (6) and (8) one has

$$\nabla \times \left[\nabla \times \nabla \times \Pi + \epsilon\mu\ddot{\Pi}\right] = 0 \tag{10}$$

which implies that the vector within the square brackets in Eq. (10) is rotation free, that is

$$\nabla \times \nabla \times \Pi + \frac{1}{c^2}\ddot{\Pi} = \nabla\psi \tag{11}$$

where ψ is an arbitrary scalar function. Selecting $\psi = \nabla \cdot \Pi$ and using the vector identity $\nabla \times \nabla \times \Pi = \nabla(\nabla \cdot \Pi) - \nabla^2\Pi$, Eq. (11) reduces to the wave equation

$$\nabla^2\Pi - \frac{1}{c^2}\ddot{\Pi} = 0 \tag{12}$$

propagating with the velocity c of light in the medium.

The solution forms (8), (9), and (12) holds for any light wave propagation. In the special case of propagation through an optical fiber, if its axis is chosen as the z-axis, and a system of cylindrical polar coordinates (r, θ, z) is employed, then the Hertz vector is oriented parallel to the axis of z having only one component:

$$\Pi = \Pi_3\,\mathbf{e}_z \tag{13}$$

where $(\mathbf{e}_r, \mathbf{e}_\theta, \mathbf{e}_z)$ are unit vectors in the coordinate system forming a basis. For this system, Eq. (12) reduces to a single wave equation

$$\frac{\partial^2\Pi_3}{\partial r^2} + \frac{1}{r}\frac{\partial\Pi_3}{\partial r} + \frac{1}{r^2}\frac{\partial^2\Pi_3}{\partial\theta^2} + \frac{\partial^2\Pi_3}{\partial z^2} - \frac{1}{c^2}\frac{\partial^2\Pi_3}{\partial t^2} = 0 \tag{14}$$

and the components $(\mathbf{E}_r, \mathbf{E}_\theta, E_z)$ and (H_r, H_θ, H_z) of E and H from Eqs. (9) and (8) are given by

$$E_r = \frac{\partial^2\Pi_3}{\partial r\partial z}, \quad E_\theta = \frac{1}{r}\frac{\partial^2\Pi_3}{\partial\theta\partial z}, \quad E_z = \frac{\partial^2\Pi_3}{\partial z^2} - \frac{1}{c^2}\frac{\partial^2\Pi_3}{\partial t^2} \tag{15}$$

$$H_r = \frac{\epsilon}{r}\frac{\partial^2\Pi_3}{\partial\theta\partial t}, \quad H_\theta = -\epsilon\frac{\partial^2\Pi_3}{\partial r\partial t}, \quad H_z = 0 \tag{16}$$

In Eq. (15), the expression for E_z has been simplified by using Eq. (14).

The solution of Eq. (14) for the component Π_3 is formally different in the two regions $r \leq a$ of the core and that of the cladding region $r > a$. If the function in the two regions is denoted by $\Pi_3^{(1)}$ and $\Pi_3^{(2)}$, respectively, satisfying Eq. (14) with velocities $c_1 = 1/\sqrt{\epsilon_1\mu_1}$ and $c_2 = 1/\sqrt{\epsilon_2\mu_2}$, then it easily follows that the modal decomposition of solution for progressive waves in the composite medium is of the form

$$\Pi_3^{(1)} = A_m J_m\left(\frac{u}{a}r\right)e^{im\theta}e^{i(kz-\omega t)}, \quad 0 \leq r \leq a \tag{17}$$

and

$$\Pi_3^{(2)} = C_m K_m\left(\frac{v}{a}r\right)e^{im\theta}e^{i(kz-\omega t)}, \quad r > a \tag{18}$$

where A_m, C_m are constant amplitudes, and $J_m(\cdot)$, $K_m(\cdot)$ are, respectively, the Bessel functions and the modified Bessel functions ([16], p. 358, 374) such that $m = 0, 1, 2, 3, \cdots$. The quantities k and ω are, respectively, the wave number and the angular frequency of the propagating wave and the two parameters u and v in Eqs. (17) and (18) are defined by the expressions

$$u = ak\sqrt{\frac{c_p^2}{c_1^2} - 1}, \quad v = ak\sqrt{1 - \frac{c_p^2}{c_2^2}} \tag{19}$$

in which $c_p = \omega/k$ is the phase velocity of propagation. Evidently, $c_1 < c_p < c_2$ for real values of u and v. It may be noted that Eq. (17) represents an oscillatory solution in the core region, while Eq. (18) represents an evanescent wave as $r \to \infty$, in the domain of the cladding.

At the interface $r = a$ of the two media, the tangential components of **E** and **H** are continuous, so that $E_z^{(1)} = E_z^{(2)}$ and $H_\theta^{(1)} = H_\theta^{(2)}$. These two conditions yield the pair of equations

$$u^2 J_m(u) A_m = -v^2 K_m(v) C_m \tag{20}$$

and

$$u J_m'(u) A_m = v K_m'(v) C_m \tag{21}$$

as $\epsilon_1 \approx \epsilon_2$. The third condition $E_\theta^{(1)} = E_\theta^{(2)}$ is satisfied identically for $m = 0$ the fundamental mode as also approximately in the higher modes $m \geq 1$ as the EM field is weaker in these modes in the very thin fiber as $J_m(ur/a) \to 0$ for $r \to 0$ and consequently $C_m \approx 0$. Thus, eliminating the constants A_m and C_m from the pair of Eqs. (20) and (21), it follows that the propagation is dispersive, satisfying the equation

$$\frac{1}{u}\frac{J_m'(u)}{J_m(u)} + \frac{1}{v}\frac{K_m'(v)}{K_m(v)} = 0 \tag{22}$$

Eq. (22) is well documented in most texts such as that of Keiser [3]. The equation was also derived by Bose [7] from the more general treatment of Stratton [8]. This equation can be rewritten by using the well-known properties of $J_m(\cdot)$ and $K_m(\cdot)$ Abramowitz and Stegun ([16], p. 361, 375) as

$$\frac{1}{u}\frac{J_{m+1}(u)}{J_m(u)} + \frac{1}{v}\frac{K_{m+1}(v)}{K_m(v)} - m\left(\frac{1}{u^2} + \frac{1}{v^2}\right) = 0 \tag{23}$$

Eq. (23) can be nondimensionalized by introducing familiar variables employed in fiber-optic literature. Let f (commonly denoted by V) be the nondimensional frequency of the propagating wave such that

$$f = \sqrt{u^2 + v^2} = a\omega\sqrt{\frac{1}{c_1^2} - \frac{1}{c_2^2}} = \frac{a\omega}{c_0}\sqrt{n_1^2 - n_2^2} \tag{24}$$

and b defined by the equation

$$b = \frac{c_0^2/c_p^2 - n_2^2}{n_1^2 - n_2^2} = \frac{v^2}{f^2} \tag{25}$$

replacing the relative phase velocity c_p/c_0, then u and v can be expressed as

$$U = \sqrt{1-bf} \quad \text{and} \quad v = \sqrt{bf} \tag{26}$$

Eq. (23) then takes up the form

$$\frac{1}{\sqrt{1-bf}} \frac{J_{m+1}\left(\sqrt{1-bf}\right)}{J_m\left(\sqrt{1-bf}\right)} + \frac{1}{\sqrt{bf}} \frac{K_{m+1}\left(\sqrt{bf}\right)}{K_m\left(\sqrt{bf}\right)} - \frac{m}{b(1-b)f^2} = 0 \tag{27}$$

Alternatively, replacing m by $-m$ without violating periodicity of the solution forms (17) and (18) with respect to θ, and using the properties of the Bessel functions $J_{-m}(u) = (-1)^m J_m(u)$ and $K_m(v) = (-1)^m K_m(v)$, Eq. (27) can also be written in the form

$$\frac{1}{\sqrt{1-bf}} \frac{J_{m-1}\left(\sqrt{1-bf}\right)}{J_m\left(\sqrt{1-bf}\right)} - \frac{1}{\sqrt{bf}} \frac{K_{m-1}\left(\sqrt{bf}\right)}{K_m\left(\sqrt{bf}\right)} - \frac{m}{b(1-b)f^2} = 0 \tag{28}$$

The two Eqs. (27) and (28) are somewhat different from those obtained by assuming the "weakly guiding approximation" Bhadra and Ghatak [17].

For a numerical study of the dispersion phenomenon, Eq. (27) is solved for typical values of $n_1 = 1.5$ and $n_2 = 1.48515$ considered by Bhadra and Ghatak. Higher modes generally speaking contribute little to a propagating pulse. Hence, for a qualitative view of the dispersion curves, the fundamental mode $m = 0$ is considered. The next mode $m = 1$ yields similar results. As $0 < b < 1$, and $f > 0$, the zeros of the left-hand side of Eq. (27) are first isolated in the domain $b = 0(0.1)1$ and $f = 0(0.25)10$. Next, the zeros are refined by a simple bisection method. The dispersion curves for the two modes are shown in **Figures 1** and **2**.

Since the phase velocity of propagation is determined by b depending on the frequency f according to Eq. (27), it means that the light waves through the core of the fiber propagate as a group of waves with certain velocity $c_g = d\omega/dk$ ([6], p. 21). This feature is observed in the experiments of Leland [18]. It is also observed in that study that the waves shoot up for certain frequencies of propagation. Such bursts occur if the group velocity c_g vanishes for some real values of f. For existence of such

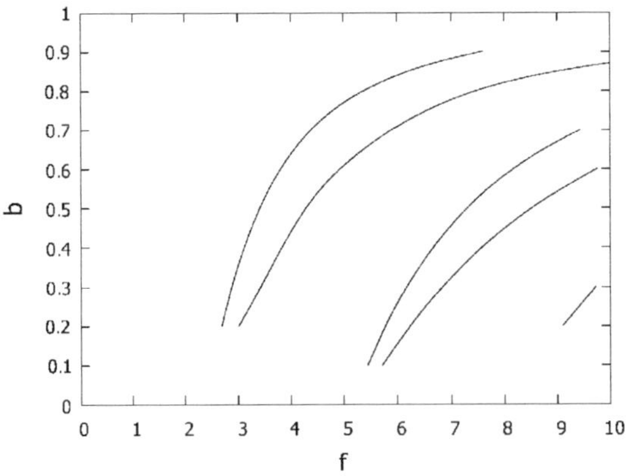

Figure 1.
Dispersion curves: Mode $m = 0$.

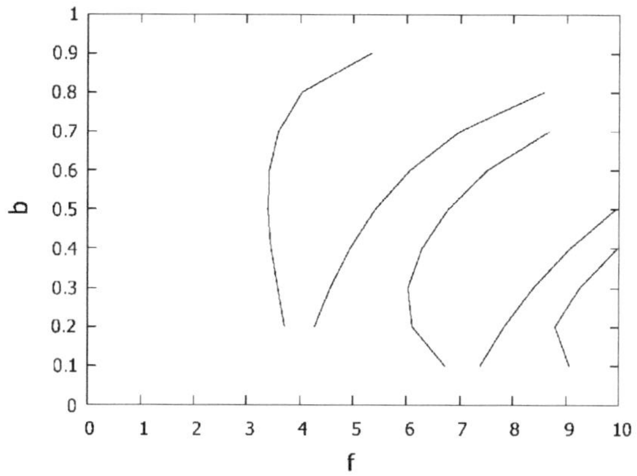

Figure 2.
Dispersion curves: Mode $m = 1$.

values, $d\omega/dk$ is set to zero by differentiating Eq. (27). Thus, for $m = 0$ one gets the equation

$$\frac{1}{u^3 J_0^2(u)}\left[u\{J_0^2(u)+J_1^2(u)\}-2J_0(u)J_1(u)\right]-\frac{1}{v^3K_0^2(v)}\left[v\{K_0^2(v)+K_1^2(v)\}-2K_0(v)K_1(v)\right]=0 \tag{29}$$

and for the case $m = 1$, the equation

$$\frac{1}{u^3J_0^2(u)}\left[uJ_1^2(u)-\{2J_1(u)+uJ_0(u)\}J_2(u)\right] - \frac{1}{v^3K_1^2(v)}\left[vK_1^2(v)-\{2K_1(v)+vK_0(v)\}K_2(v)\right]-\frac{2}{(1-b)v^3}=0 \tag{30}$$

The roots of Eqs. (29) and (30) follow a certain trend for different values of f for the numerical data used earlier. These are shown in **Figures 3** and **4**, respectively, for the two modes considered above. The vanishing of the group velocity for certain values of f provides an explanation for bursts in transmission as stated earlier. This

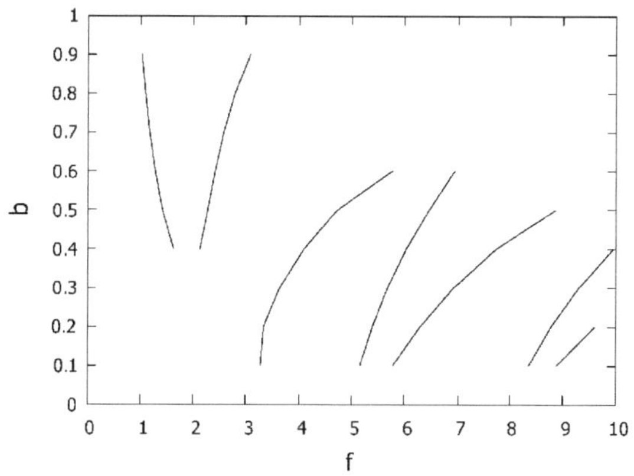

Figure 3.
Zero group velocity curves: Mode: $m = 0$.

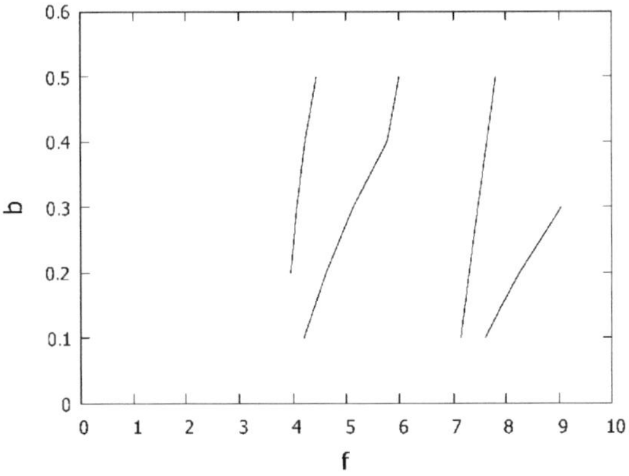

Figure 4.
Zero group velocity curves: Mode: $m = 1$.

theory was reported by the present author [7], who proceeded to develop a theory of the exponential pileup of data packets in such propagation.

3. Macrobending power losses

Consider now a gradually bent optical fiber with small curvatures, as shown in **Figure 5**. It is assumed here for simplicity of analysis that the curved cable lies in a plane, which is conveniently considered to be the (z, y) plane. Here, the focus is on the curvature of bends, neglecting the secondary effects of elastic strain, heating, and even microcracking of the core material. For analysis of the effect of the geometrical "macrobending" effect, intrinsic coordinates (s, ψ) for the curve are employed, as shown in **Figure 5**. The Cartesian representation is thus given in parametric form as:

$$z = f_1(s), y = f_2(s), \quad \text{with} \quad \tan\psi = y' \tag{31}$$

where the prime denotes differentiation with respect to z. For propagation of light waves along the increasing arc length s, the Hertz vector Π is largely oriented in that direction, but has a transversal component as well because of the curvature. Thus, it can be represented as

$$\Pi = \Pi_s \, \mathbf{e}_s + \Pi_\psi \, \mathbf{e}_\psi \tag{32}$$

where $\mathbf{e}_s, \mathbf{e}_\psi$ are unit vectors along the increasing s and ψ directions. These unit vectors can be expressed in terms of $(\mathbf{e}_z, \mathbf{e}_y)$ in the z and y directions, respectively, by the equations

Figure 5.
Intrinsic coordinate system (s, ψ) of a point P of a bent fiber cable.

$$\mathbf{e}_s = \cos\psi\,\mathbf{e}_z + \sin\psi\,\mathbf{e}_y, \quad \mathbf{e}_\psi = -\sin\psi\,\mathbf{e}_z + \cos\psi\,\mathbf{e}_y \tag{33}$$

or, reciprocally

$$\mathbf{e}_z = \cos\psi\,\mathbf{e}_s - \sin\psi\,\mathbf{e}_\psi, \quad \mathbf{e}_y = \sin\psi\,\mathbf{e}_s + \cos\psi\,\mathbf{e}_\psi \tag{34}$$

Inserting the transformation (33) in (32), the wave equation for Π viz. Eq. (12) results in the pair of equations

$$\begin{aligned}\nabla^2\Pi_s - 2\left(\frac{\partial\psi}{\partial z}\frac{\partial\Pi_\psi}{\partial z} + \frac{\partial\psi}{\partial y}\frac{\Pi_\psi}{\partial y}\right) - \frac{1}{c^2}\ddot{\Pi}_s + \Pi_s\{\cos\psi\nabla^2(\cos\psi) + \sin\psi\nabla^2(\sin\psi)\}\\ +\Pi_\psi\{\cos\psi\nabla^2(\sin\psi) + \sin\psi\nabla^2(\cos\psi)\} = 0\end{aligned} \tag{35}$$

$$\begin{aligned}\nabla^2\Pi_\psi = 2\left(\frac{\partial\psi}{\partial z}\frac{\partial\Pi_s}{\partial z} + \frac{\partial\psi}{\partial y}\frac{\partial\Pi_s}{\partial y}\right) - \frac{1}{c^2}\ddot{\Pi}_\psi + \Pi_s\{\cos\psi\nabla^2(\sin\psi) - \sin\psi\nabla^2(\cos\psi\}\\ +\Pi_\psi\{\cos\psi\nabla^2(\cos\psi) - \sin\psi\nabla^2(\sin\psi)\} = 0\end{aligned} \tag{36}$$

The two coupled Eqs. (35) and (36) are linear but with complicated coefficients depending on the nature of the curved form of the cable. However, if the curvature of the cable is small all along its length, then the angle ψ is small, so that $\sin\psi\approx 0$, $\cos\psi\approx 1$, $\partial/\partial z(\cdot)\approx\partial/\partial s(\cdot)$, $\partial/\partial y(\cdot)\approx 0$, and $\partial\psi/\partial z\approx\partial\psi/\partial s \equiv \kappa(s)$ = curvature. Thus for this case, the coefficients of Π_s and Π_ψ in the above two equations become negligible, and therefore, approximately the two equations simplify as

$$\nabla^2\Pi_s - \frac{1}{c^2}\ddot{\Pi}_s = 2\kappa(s)\frac{\partial\Pi_\psi}{\partial s} \tag{37}$$

and

$$\nabla^2\Pi_\psi - \frac{1}{c^2}\ddot{\Pi}_\psi = -2\kappa(s)\frac{\partial\Pi_s}{\partial s} \tag{38}$$

The complementary functions of the general solution of Eqs. (37) and (38) for propagating waves in the s direction are

$$\Pi_s = C_1 e^{i(ks-\omega t)}, \quad \Pi_\psi = C_2 e^{i(ks-\omega t)}, \quad C_1, C_2 = \text{Constant} \tag{39}$$

while the particular solutions, taking into account the right-hand sides of Eqs. (37) and (38), are small on account of the small curvature of the curved cable. Hence, $\Pi_s/\Pi_\psi\approx C_1/C_2$—a constant, say $\epsilon/2$. The constant ε will be positive if the curvature is positive, otherwise negative as Π_ψ will have a negative orientation, the principal equation of propagation in the s direction, viz. Eq. (37), can be taken to be of the form

$$\nabla^2\Pi_s - \frac{1}{c^2}\ddot{\Pi}_s = \epsilon\,|\kappa(s)|\frac{\partial\Pi_s}{\partial s} \tag{40}$$

where $\epsilon > 0$ is a small parameter. Let the modal solution of Eq. (40) with the variable cofficient $|\kappa(s)|$ be

$$\Pi_s = f_1(r)g_1(s)e^{im\theta}e^{-i\omega t} \tag{41}$$

then substituting Eq. (41) in Eq. (40), and separating the variables r and s, one gets

$$\frac{1}{f_1}\left[f_1''(r)+\frac{1}{r}f_1'(r)+\left(\frac{\omega^2}{c^2}-\frac{m^2}{r^2}\right)f_1(r)\right]=-\frac{1}{g_1(s)}\left[g_1''(s)-\epsilon|\kappa(s)|g_1'(s)\right]=k^2 \quad (42)$$

where k is a constant in virtue of the fact that the two sides of the first equality are of independent variables r and s. It immediately follows that $f_1(r)$ is a Bessel function, while $g_1(s)$ is a solution of the linear equation whose right-hand side contains the small multiplier constant ϵ:

$$g_1''(s)+k^2 g_1(s)=\epsilon|\kappa(s)|g_1'(s)\equiv\epsilon F(g_1'(s)),\quad \text{(say)} \quad (43)$$

As ε is small, an averaged solution of Eq. (43) is sought following the method of Krylov and Bogoliubov ([19], p. 165). For that purpose, the general solution of Eq. (43) is considered to be of the form

$$g_1(s)=a\,e^{i\phi(s)},\quad \phi(s)=ks+\eta(s) \quad (44)$$

with a slowly varying phase $\eta(s)$. The form (44) is suggested by setting $\epsilon=0$ in Eq. (43), leading to a simple harmonic oscillator of wave number k with a constant phase. Substituting the formal solution (44) in Eq. (43), it follows that

$$2\eta'(s)=-\frac{\varepsilon}{a}e^{-i\phi}F\left[aie^{i\phi}(\omega_0+\eta'(s))\right]+\frac{i}{2}\frac{d\{\eta'(s)\}^2}{d\eta} \quad (45)$$

Inasmuch as η is a slowly varying function of s, $\eta'(s)$ is small, so that to the first order of smallness

$$2k\eta'(s)=-\frac{\epsilon}{a}e^{-i\phi}F\left[akie^{i\phi}\right] \quad (46)$$

Now, the function on the right-hand side of Eq. (46) is periodic with slowly varying period 2π. So replacing the function by its average, one gets approximately

$$\eta'(s)=-\frac{\epsilon}{2ak}\frac{1}{2\pi}\int_0^{2\pi}e^{-i\phi}F(akie^{i\phi})\,d\phi \quad (47)$$

Thus, using the definition of $F(s)$, and integrating with respect to s, one obtains to the first order in ϵ

$$\eta(s)=-\frac{i\epsilon s}{4\pi}\int_0^{2\pi}\left|\kappa\left(\frac{\phi}{k}\right)\right|d\phi \quad (48)$$

and therefore from Eq. (44), it follows that

$$g_1(s)=a\ \exp\left[\left\{ik+\frac{\epsilon}{4\pi}\int_0^{2\pi}\left|\kappa\left(\frac{\phi}{k}\right)\right|d\phi\right\}s\right] \quad (49)$$

When the form (49) of $g_1(s)$ is inserted in the general solution (41), it is at once becomes apparent that the positive real exponential part of the solution for increasing s indicates that attenuation of light waves occurs, proportional to the curvature of bending resulting in reduction of power of any signal sent through a bent cable.

A detailed analysis of the problem was earlier carried out for a circularly bent cable by Bose [12], using simple toroidal coordinates to show that the macrobending loss of power is small when the radius R of the bend exceeds a critical value R_c, the loss is proportional to the winding angle, or in other words, proportional to the length of the circularly bent part. When R is smaller than R_c, the power loss was however found to be much greater—proportional to the cube of the winding angle, that is to say proportional to the cube of the length of the cable following a lengthy analysis. The present finding of power loss in macrobending is consistent with the first part of that finding for a circular bend.

4. Graded-index fibers

In this case, the refractive index of the core material is variable in the radial r direction with a given profile $n(r)$, so that if ϵ_1, μ_1 are the electric permittivity and the magnetic permeability

$$\epsilon_1\mu_1 = \frac{n^2}{c_0^2}, \quad 0 \le r \le a \tag{50}$$

where c_0 is the velocity of light in the vacuum. Similarly, for the cladding material if ϵ_2, μ_2 are respective permittivity and permeability, then

$$\epsilon_2\mu_2 = \frac{n_2^2}{c_0^2}, \quad r \ge a \tag{51}$$

where n_2 denotes the constant refractive index of the cladding material. The variability of n in Eq. (50) is due to variability of the permittivity ϵ_1 alone, whereas μ_1 is regarded as invariable in the dielectric medium ([1], p. 11). Maxwell Eqs. (1)–(4), together with Eqs. (5) and (6), still hold, but due to variability of ϵ_1, the definition of the Hertz vector Π is slightly modified following Mohsen [20]. Following that procedure, let

$$\mathbf{B} = \nabla \times \frac{\partial \Pi}{\partial t}, \quad \mathbf{H} = \frac{1}{\mu}\nabla \times \frac{\partial \Pi}{\partial t} \tag{52}$$

instead of Eq. (7). Hence, by Eqs. (4) and (5), since μ is not a variable

$$\epsilon\frac{\partial \mathbf{E}}{\partial t} = \frac{1}{\mu}\nabla \times \nabla \times \frac{\partial \Pi}{\partial t} \tag{53}$$

or, by integration with respect to time t,

$$\mathbf{E} = \frac{1}{\mu\epsilon}\nabla \times \nabla \times \Pi \tag{54}$$

and so the displacement current given by Eq. (5) yields

$$\mathbf{D} = \frac{1}{\mu}\nabla \times \nabla \times \Pi \tag{55}$$

automatically satisfying Eq. (1). Thus, all the field quantities are determined in terms of Π. To find the equation governing Π, one has from Eqs. (3) and (52),

$$\nabla \times \left(\frac{1}{\mu\epsilon} \nabla \times \nabla \times \Pi \right) = -\nabla \times \frac{\partial^2 \Pi}{\partial t^2} \tag{56}$$

which is satisfied if

$$\frac{1}{\mu\varepsilon} \nabla \times \nabla \times \Pi + \frac{\partial^2 \Pi}{\partial t^2} = \nabla\psi \tag{57}$$

where ψ is an arbitrary scalar field. Hence, using the well-known vector identity for curl,

$$\nabla(\nabla \cdot \Pi) - \nabla^2 \Pi + \mu\epsilon \frac{\partial^2 \Pi}{\partial t^2} = \mu\epsilon\nabla\psi \tag{58}$$

or,

$$\mu\epsilon \frac{\partial^2 \Pi}{\partial t^2} - \nabla^2 \Pi = \mu[\nabla(\epsilon\psi) - \psi\nabla\epsilon] - \nabla(\nabla \cdot \Pi) \tag{59}$$

Now since ψ is arbitrary, choosing

$$\psi = \frac{1}{\mu\epsilon} \nabla \cdot \Pi \tag{60}$$

Eq. (59) reduces to the wave equation for the varying index medium as

$$\nabla^2 \Pi + \frac{\nabla\epsilon}{\epsilon} (\nabla \cdot \Pi) = \mu\epsilon \frac{\partial^2 \Pi}{\partial t^2} \tag{61}$$

Evidently, for a homogeneous medium, such as the cladding material, $\nabla\epsilon = 0$, so that Eq. (61) reduces to a wave equation propagating with a velocity $1/\sqrt{\mu\epsilon}$.

In the case of a light wave propagating along the axis of the cable, there is no transversal component of the Hertz vector, so that if $\Pi_1^{(1)}$ and $\Pi_3^{(1)}$ are the radial and longitudinal components of $\Pi^{(1)}$ in the core

$$\Pi^{(1)} = \Pi_1^{(1)} \mathbf{e}_1 + \Pi_3^{(1)} \mathbf{e}_3 \tag{62}$$

where $\mathbf{e}_1$ and $\mathbf{e}_3$ are unit vectors in the radial and longitudinal directions. Thus, Eq. (61) splits into the scalar equations

$$\begin{aligned} \nabla^2 \Pi_1^{(1)} + \frac{1}{\epsilon_1} \frac{\partial \epsilon_1}{\partial r} \frac{1}{r} \frac{\partial \left(r\Pi_1^{(1)} \right)}{\partial r} - \mu_1\epsilon_1 \frac{\nabla^2 \Pi_1^{(1)}}{\partial t^2} \\ = -\frac{1}{\epsilon_1} \frac{\partial \epsilon_1}{\partial r} \frac{\partial \Pi_3^{(1)}}{\partial z} \end{aligned} \tag{63}$$

and

$$\nabla^2 \Pi_3^{(1)} = \mu_1\epsilon_1 \frac{\nabla^2 \Pi_3^{(1)}}{\partial t^2} \tag{64}$$

It is to be noted that while Eq. (64) is a standard form wave equation with variable propagation velocity $1/\sqrt{\mu_1 \epsilon_1}$, Eq. (21) is a modified form with a forcing function due to the axial component $\Pi_3^{(1)}$.

The corresponding electric and magnetic intensities in the core following Eqs. (54) and (52) are thus

$$E_r^{(1)} = \frac{1}{\mu_1 \epsilon_1}\left[-\frac{1}{r^2}\frac{\partial^2 \Pi_1^{(1)}}{\partial \theta^2} - \frac{\partial^2 \Pi_1^{(1)}}{\partial z^2} + \frac{\partial^2 \Pi_3}{\partial z \partial r}\right] \tag{65}$$

$$E_\theta^{(1)} = \frac{1}{\mu \epsilon_1}\left[\frac{\partial}{\partial r}\left(\frac{1}{r}\frac{\partial \Pi_1^{(1)}}{\partial \theta} + \frac{1}{r}\frac{\partial^2 \Pi_3^{(1)}}{\partial \theta \partial z}\right.\right] \tag{66}$$

$$E_z^{(1)} = \frac{1}{\mu_1 \epsilon_1}\left[\frac{1}{r}\frac{\partial}{\partial r}\left\{r\left(\frac{\partial \Pi_1^{(1)}}{\partial z} - \frac{\partial \Pi_3^{(1)}}{\partial r}\right)\right\} - \frac{1}{r^2}\frac{\partial^2 \Pi_3^{(1)}}{\partial \theta^2}\right] \tag{67}$$

$$H_r^{(1)} = \frac{1}{\mu_1}\frac{\partial}{\partial \theta}\left(\frac{\partial \Pi_3^{(1)}}{\partial t}\right) \tag{68}$$

$$H_\theta^{(1)} = -\frac{1}{\mu_1}\left[\frac{\partial}{\partial r}\left(\frac{\partial \Pi_3^{(1)}}{\partial t}\right) - \frac{\partial}{\partial z}\left(\frac{\partial \Pi_1^{(1)}}{\partial t}\right)\right] \tag{69}$$

$$H_z^{(1)} = -\frac{1}{\mu_1}\frac{\partial}{r \partial \theta}\left(\frac{\partial \Pi_1^{(1)}}{\partial t}\right) \tag{70}$$

In the cladding material, which is homogeneous, the radial component of the Hertz vector $\Pi_1^{(2)}$ vanishes, so that the electric and magnetic field intensities in the medium are given by the single wave equation for $\Pi_3^{(2)}$ with constant wave velocity of propagation $1/\sqrt{\mu_2 \epsilon_2}$. Hence, in this case the field intensities $E_r^{(2)}$. $E_\theta^{(2)}$, $E_z^{(2)}$, $H_r^{(2)}$, $H_\theta^{(2)}$, and $H_z^{(2)}$ reduce to the same form as Eqs. (65)–(70) with $\Pi_1^{(2)}$ set to zero and μ_1, ϵ_1 replaced by μ_2, ϵ_2, respectively. It is to be noted that in this case $H_z^{(2)} = 0$.

The formalism developed above is now applied to the parabolic-index case. Here, the index of refraction n of the core material is assumed to be parabolic of degree α such that

$$n^2 = n_1^2\left[1 - 2\Delta\left(\frac{r}{a}\right)^\alpha\right], \quad 0 \le r \le a \tag{71}$$

where

$$\Delta = \frac{n_1^2 - n_2^2}{2n_1^2} < < 1 \tag{72}$$

in which the index n_1 of the core is slightly greater than n_2 ([3], p. 68). In this case, using Eq. (71), one has

$$\frac{1}{\epsilon_1}\frac{d\epsilon_1}{dr} = -\frac{2\alpha\Delta}{a}\left(\frac{r}{a}\right)^{\alpha-1} \tag{73}$$

This coefficient is of order Δ appearing in Eq. (63). However, it is generally neglected altogether on physical ground ([1], p. 11), but is retained here without much difficulty.

The propagation of light waves in an optical fiber can take place in different modes as in the case of uniform homogeneous core step-index cable, essentially because of smallness of Δ. In such propagation, the modal solution of Eq. (64) is of the form

$$\Pi_3^{(1)} = f_3(r) e^{im\theta} e^{i(kz-\omega t)}, \quad 0 \le r \le a \tag{74}$$

where $f_3(r)$ is a function of the radial coordinate r only, $i = \sqrt{-1}$, m the mode, k the wave number, and ω the circular frequency of the propagating wave. Inserting Eq. (74) in Eq. (64) and using the profile of the refractive index given by Eq. (71), one gets the equation

$$f_3''(r) + \frac{1}{r} f_3'(r) - \frac{m^2}{r^2} f_3(r) + \left[\frac{\omega^2}{c_0^2} n_1^2 \left\{1 - 2\Delta\left(\frac{r}{a}\right)^{\alpha}\right\} - k^2\right] f_3(r) = 0 \tag{75}$$

where the primes denote differentiation with respect to the argument. Rearranging the equation for a perturbation solution,

$$r^2 f_3''(r) + r f_3'(r) + \left[\left(\frac{\omega^2}{c_0^2} n_1^2 - k^2\right) r^2 - m^2\right] f_3(r) = \frac{2\omega^2 n_1^2 \Delta}{c_0^2} r^2 \left(\frac{r}{a}\right)^{\alpha} f_3(r) \tag{76}$$

It is now convenient to rescale the independent variable r by the nondimensional variable r' defined by the relation

$$r = \left(\frac{a}{u}\right) r_1 \tag{77}$$

in which u is a nondimensional parameter defined by Eq. (19), or

$$u = ak\sqrt{\frac{c_p^2 n_1^2}{c_0^2} - 1} \tag{78}$$

Thus, Eq. (76) transforms into the equation

$$r_1^2 f_3''(r_1) + r_1 f_3'(r_1) + \left(r_1^2 - m^2\right) f_3(r_1) = B_1 r_1^{(\alpha+2)} f_3(r_1) \tag{79}$$

where the nondimensional constant B_1 is defined as

$$B_1 = \frac{2c_p^2 a^2 k^2 n_1^2}{c_0^2 u^{\alpha+2}} \tag{80}$$

If the right-hand side of Eq. (79) is set to zero, the solution is $f_3(r_1) = AJ_m(r_1)$, where A is a constant and $J_m(\cdot)$ is the Bessel function of order m. Hence, inserting this first approximation into the right-hand side of Eq. (79), the general solution of that equation can be assumed to be of the form

$$f_3(r_1) = A\left[J_m(r_1) + B_1 \Delta r_1^{\lambda}\left(C_0 + C_1 r_1^2 + C_2 r_1^4 + \cdots\right)\right] \tag{81}$$

where the coefficients $C_0, C_1, C_2 \cdots$ are certain constants to be determined. Inserting the expression for $f_3(r_1)$ in Eq. (79) with $f_3(r_1)$ replaced by $AJ_m(r_1)$ on its right-hand side, one obtains

$$r_1^{\lambda}\left[(\lambda^2 - m^2)C_0 + \left\{\left[(\lambda+2)^2 - m^2\right]C_1 + C_0\right\}r_1^2 + \left\{\left[(\lambda+4)^2 - m^2\right\}C_2 + C_1\right\}r_1^4 + \cdots\right]$$
$$= r_1^{\alpha+2}J_m(r_1) = \frac{r_1^{\alpha+m+2}}{2^m}\sum_{\nu=0}^{\infty}\frac{(-1)^{\nu}\left(r_1^2/4\right)^{\nu}}{\nu!(m+\nu)!} \tag{82}$$

Hence, equating coefficients of the powers of r_1 on the two sides of Eq. (82), one obtains

$$\lambda = \alpha + m + 2 \tag{83}$$

$$C_0 = \frac{1}{2^m\, m!}\frac{1}{\lambda^2 - m^2} \tag{84}$$

with the recurrence relation for the determination of the coefficients C_ν as

$$\left[(\lambda + 2\nu)^2 - m^2\right]C_\nu + C_{\nu-1} = \frac{1}{2^m}\frac{(-1)^{\nu}}{\nu!(m+\nu)!4^{\nu}}, \quad \nu = 1, 2, 3, \cdots \tag{85}$$

yielding all the coefficients $C_0, C_1, C_2, \cdots$ of the solution (81).

For a complete description of the electromagnetic field, the radial component $\Pi_1^{(1)}$, using Eq. (63), is obtained from the solution of the equation

$$\nabla^2\Pi_1^{(1)} - \mu_1\epsilon_1\frac{\partial^2\Pi_1^{(1)}}{\partial t^2} = \frac{2\alpha\Delta}{a}\left(\frac{r}{a}\right)^{\alpha-1}\left[\frac{\partial\Pi_3^{(1)}}{\partial z} + \frac{1}{r}\frac{\partial}{\partial r}\left(r\Pi_1^{(1)}\right)\right] \tag{86}$$

in which $\Pi_3^{(1)}$ is given by Eq. (74). The solution of Eq. (86) is of the modal form

$$\Pi_3^{(1)} = f_1(r)e^{im\theta}e^{i(kz-\omega t)} \tag{87}$$

where the function $f_1(r)$ satisfies the equation to the first order of Δ as,

$$r^2 f_1''(r) + rf_1'(r) + \left(\frac{u^2}{a^2}r^2 - m^2\right)f_1(r) = \frac{2\alpha\Delta}{a}r^2\left(\frac{r}{a}\right)^{\alpha-1}ikAJ_m\left(\frac{a}{u}r\right) \tag{88}$$

or, resorting to the independent variable r_1 defined by Eq. (77), the Eq. (88) becomes

$$r_1^2 f_1''(r_1) + r_1 f_1'(r_1) + (r_1^2 - m^2)f_1(r_1) = B_2\Delta A r_1^{(\alpha+1)}J_m(r_1) \tag{89}$$

where the nondimensional constant B_2 is given by the equation

$$B_2 = \frac{2iak\alpha}{u^{\alpha+1}} \tag{90}$$

If the Bessel function $J_m(r_1)$ is expanded as a power series, the solution of Eq. (89) is of the form

$$f_1(r_1) = B_2 \Delta A \left(D_0 + D_1 r_1^2 + D_2 r_2^4 + \cdots\right) \tag{91}$$

where $D_0, D_1, D_2, \cdots$ are certain constants. Inserting the form (91) in Eq. (89) and equating the coefficients of the different powers of r_1, it again follows that

$$\mu = \alpha + m + 1 \tag{92}$$

$$D_0 = \frac{1}{2^m m!}\frac{1}{\mu^2 - m^2} \tag{93}$$

and the recurrence relation for the rest of the coefficients D_ν as

$$\left[(\mu + 2\nu)^2 - m^2\right] D_\nu + D_{\nu-1} = \frac{1}{2^m}\frac{(-1)^\nu}{\nu!(m+\nu)!4^\nu}, \quad \nu = 1, 2, 3, \cdots \tag{94}$$

yielding the values of $D_0, D_1, D_2, \cdots$ in the solution of Eq. (91). Using the forms (74), (81) and (87), (91) for the field components of the Hertz vector $\Pi_3^{(1)}$ and $\Pi_1^{(1)}$, respectively, in Eqs. (65)–(70), the expressions for the electromagnetic field components within the fiber core can be explicitly written down. The required components for the satisfaction of the components are:

$$E_z^{(1)} = \frac{c_0^2}{n_1^2}\left(1 + \frac{2\Delta}{u^\alpha} s^\alpha\right) \frac{uA}{a^2}\left[u\left\{J_m(r_1) - B_1 \Delta r_1^{\lambda-2}\left[C_0\left(\lambda^2 - m^2\right) + C_1\left\{(\lambda+2)^2 - m^2\right\} + C_2\left\{(\lambda+4)^2 - m^2\right\} r_1^4 + \cdots\right]\right\} + iakB_2\Delta s^{\mu-1}\left\{D_0(\mu+1) + D_1(\mu+3)s^2 + D_2(\mu+5)s^4 + \cdots\right\}\right] \tag{95}$$

and

$$H_\theta^{(1)} = \frac{i\omega u A}{a\mu_1}\Big[J'_m(s) + B_1 \Delta s^{\lambda-1}\left\{\lambda C_0 + (\lambda+2) C_1 s^2 + (\lambda+4) C_2 s^4 + \cdots\right\} - \frac{iak}{u} B_2 \Delta s^\mu \left(D_0 + D_1 s^2 + D_2 s^4 + \cdots\right)\Big] \tag{96}$$

In the extended medium of the cladding of the cable $r \geq a$, which is supposed to be of uniform index n_2 $(< n_1)$, the Hertz vector $\Pi_3^{(2)}$, following Eq. (64), satisfies the wave equation

$$\nabla^2 \Pi^{(2)} = \mu_2 \epsilon_2 \frac{\nabla^2 \Pi^{(2)}}{\partial t^2} \tag{97}$$

where μ_2 and ϵ_2 are, respectively, the values of the permeability and permittivity of the medium. Inasmuch as the electromagnetic field is totally axial in the medium with no radial component, $\Pi^{(2)} = \Pi_3^{(2)} \mathbf{e}_3$, where $\Pi_3^{(2))}$ satisfies an equation of the type (97). Since the electromagnetic waves diverge from the axis of the cable, the solution for $\Pi_3^{(2)}$ is of the form

$$\Pi_3^{(2)} = C K_m\left(\frac{v}{a} r\right) e^{im\theta} e^{i(kz - \omega t)} \tag{98}$$

[7], where $K_m(\cdot)$ is the modified Bessel function of order m of the second kind, and v is defined by

$$v = ak\sqrt{1 - \frac{c_p^2 n_2^2}{c_0^2}} \tag{99}$$

As in the case of the core of the cable, the electromagnetic components $E_z^{(2)}$ and $H_\theta^{(2)}$ are given by the expressions

$$E_z^{(2)} = -\frac{n_2^2}{c_0^2}\left[\frac{1}{r}\frac{\partial}{\partial r}\left(r\frac{\partial \Pi_3^{(2)}}{\partial r}\right) + \frac{1}{r^2}\frac{\partial^2 \Pi_3^{(2)}}{\partial \theta^2}\right] = -\frac{n_2^2}{c_0^2}\frac{v^2}{a^2} C K_m\left(\frac{v}{a}r\right) e^{im\theta} e^{i(kz-\omega t)} \tag{100}$$

and

$$H_\theta^{(2)} = -\frac{1}{\mu_2}\frac{\partial}{\partial r}\left(\frac{\partial \Pi_3^{(2)}}{\partial t}\right) = \frac{i\omega v}{a\mu_2} C K'_m\left(\frac{v}{a}r\right) e^{im\theta} e^{i(kz-\omega t)} \tag{101}$$

4.1 The dispersion equation

The boundary conditions at the core and the cladding lead to the dispersion equation. At the interface of the two media, the tangential components of the electric and the magnetic intensities are equal, that is to say, $E_z^{(1)} = E_z^{(2)}$, and $H_\theta^{(1)} = H_\theta^{(2)}$ at $r = a$ or from Eq. (77), $r_1 = u$. In satisfying these conditions, much simplification occurs in the equation by assuming $n_2 \approx n_1$, and $\mu_2 \approx \mu_1$ as the parameters hardly differ for the two media. In this way, by eliminating the coefficients A and C, the dispersion equation is obtained as

$$\begin{aligned}
&(1+2\Delta)\left[\left\{J_m(u) - B_1\Delta u^{\lambda-2}\left[C_0(\lambda^2 - m^2) + C_1\left\{(\lambda+2)^2 - m^2\right\}u^2 + C_2\left\{(\lambda+4)^2 - m^2\right\}u^4 + \cdots\right]\right\}\right. \\
&\quad + iakB_2\Delta u^{\mu-1}\left[D_0(\mu+1) + D_1(\mu+3)u^2 + D_2(\mu+5)u^4 + \cdots\right]\big] \times K'_m(v) \\
&\quad + u\left[J'_m(u) + B_1\Delta u^{\lambda-1}\left\{\lambda C_0 + (\lambda+2)C_1u^2 + (\lambda+4)C_2u^4 + \cdots\right\}\right. \\
&\quad \left. - iakB_2\Delta u^{\mu-1}\left(D_0 + D_1u^2 + D_2u^4 + \cdots\right)\right] \times v^2 K_m(v) = 0
\end{aligned} \tag{102}$$

Eq. (102) can be solved numerically for prescribed data. As the representative example considered in Section 2, let $n_1 = 1.5$, $n_2 = 1.48515$, and let $\alpha = 2$ (second-degree parabolic profile). The equation can then be expressed in terms of the familiar numerical frequency f and the normalized propagation constant b defined in Eqs. (24), (25), and (26) with the coefficients B_1 and B_2 given by Eqs. (87) and (90) where

$$ak = \sqrt{\frac{n_1^2 v^2 + n_2^2 u^2}{2n_1(n_1 - n_2)}} \tag{103}$$

in which u and v are as defined in Eqs. (78) and (99). Thus expressed in terms of f and b, the zeros of the left-hand side of Eq. (102) are first isolated as stated in Section 2 and then refined yielding the desired dispersion curves. The computations for the

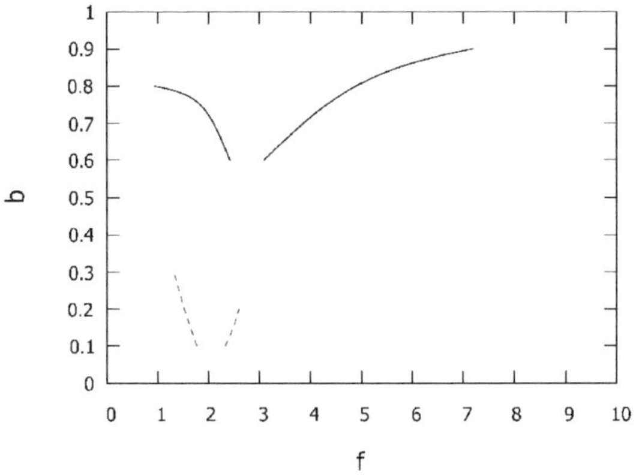

Figure 6.
Dispersion curve: Mode m = 0: black; Mode m = 1: dashed red.

modes $m = 0$ and $m = 1$ are shown together in **Figure 6**. Remarkably, the curves bifurcate into two segments over limited ranges of frequency. For the fundamental mode $m = 0$, the ranges of dispersion are more pronounced compared to the other case of $m = 1$. A further computation for the next mode $m = 2$ shows further reduction in dispersion.

5. Multimode propagation in general profile core

The procedure developed in the preceding section is now generalized for general form of the refractive index of the core medium. For that purpose, let the refractive index be represented as

$$\begin{aligned} n^2 &= n_1^2\left[1 - 2\Delta\phi\left(\frac{r}{a}\right)\right], \quad 0 \leq r \leq a \\ &= n_2^2, \quad r \geq a \end{aligned} \tag{104}$$

where

$$\Delta = \frac{n_1^2 - n_2^2}{2n_1^2\phi(1)} < < 1 \tag{105}$$

in which the index n_1 of the core is slightly greater than n_2. In all practical cases, the function $\phi(\cdot)$ is of bounded variation in $(0, 1)$, so that $\int_0^1 \sqrt{t}\phi(t)dt$ exists, and it is possible to expand $\phi(r/a)$ in a Dini series

$$\phi\left(\frac{r}{a}\right) = \sum_{\nu=0}^{\infty} b_\nu J_0\left(\lambda_\nu \frac{r}{a}\right), \quad \phi(1) = \sum_{n=0}^{\infty} b_\nu J_0(\lambda_\nu) \tag{106}$$

([21], p. 577, 600), where $\lambda_0, \lambda_1, \lambda_2, \cdots$ are the zeros of $J_1(x)$ in ascending order of magnitude, and the coefficients b_ν are given as

$$b_\nu = \frac{2}{J_0^2(\lambda_\nu)} \int_0^1 tf(t) J_0(\lambda_\nu t)\, dt \tag{107}$$

For the point $r = 0$, it at once follows that

$$\sum_{\nu=0}^{\infty} b_\nu = \phi(0) \tag{108}$$

The values of $\lambda_\nu\,(\nu = 1, 2, 3, \cdots)$ are tabulated in Abramowitz and Stegun ([16], p. 409). The value of λ_0 is evidently 0.

For modal propagation in the core medium, the Hertz vector component

$$\Pi_3^{(1)} = f_3^{(g)}(r) e^{im\theta} e^{i(kz-\omega t)}, \quad 0 \le r \le a \tag{109}$$

where $f_3^{(g)}(\cdot)$ is of the form

$$f_3^{(g)}(r_1) = A\left[J_m(r_1) + B_1' \Delta r_1^{\lambda'}\left(C_0' + C_1' r_1^2 + C_2' r_1^4 \cdots\right)\right] \tag{110}$$

and

$$B_1' = 2c_p^2 a^2 k^2 n_1^2 / c_0^2 u^2 \tag{111}$$

Inserting the form (110) in Eq. (64), one then gets

$$r_1^{\lambda'}\left[\left(\lambda'^2 - m^2\right) C_0' + \left\{\left[(\lambda'+2)^2 - m^2\right] C_1' + C_0'\right\} r_1^2 + \left\{\left[\lambda'+4)^2 - m^2\right] C_2' + C_1'\right\} r_1^4 + \cdots\right] = s^2 J_m(r_1) \sum_{\nu=0}^{\infty} b_\nu J_0\left(\frac{\lambda_\nu r_1}{u}\right) \tag{112}$$

where $r = (a/u) r_1$. (Eq.(112)) Now, the product of the two Bessel functions appearing in Eq. (112) is given by the power series expansion

$$J_m(r_1) J_0\left(\frac{\lambda_\nu r_1}{u}\right) = \left(\frac{r_1}{2}\right)^m \sum_{k=0}^{\infty} \frac{(-1)^k r_1^{2k} C_k^{(\nu)}}{4^k (k!)^2} \tag{113}$$

where

$$C_k^{(\nu)} = \begin{cases} \dfrac{k!}{(k+m)!} F\left(-k, -k-m, 1; \lambda_\nu^2/u^2\right), & \text{if} \quad \lambda_\nu < u \\ \dfrac{1}{m!}\left(\dfrac{\lambda_\nu}{u}\right)^{2k} F\left(-k, -k, m+1; u^2/\lambda_\nu^2\right), & \text{if} \quad \lambda_\nu > u \end{cases} \tag{114}$$

Inasmuch as the first argument of the hypergeometric functions $F(\cdot)$ is $-k$, the functions reduce to polynomials of degree k in the last argument λ_ν^2/u^2 or u^2/λ_ν^2 ([22], p. 960), where $F(\cdot)$ is the hypergeometric function defined by the well-known power series expansion given in Gradshteyn and Ryzhik ([22], p. 1089). The right-hand side of Eq. (112) is then of the form

$$\frac{r_1^{m+2}}{2^m} \sum_{k=0}^{\infty} C''_k r_1^{2k} \tag{115}$$

where

$$C''_k = \frac{(-1)^k}{4^k (k!)^2} \sum_{\nu=0}^{\infty} b_\nu C_k^{(\nu)} \tag{116}$$

Inserting expressions (113) and (115), (116) in Eq. (112) and equating the coefficients of different powers of r_1, one has

$$\lambda' = m + 2 \tag{117}$$

$$C'_0 = \frac{C''_0}{2^m} \frac{1}{\lambda'^2 - m^2} \tag{118}$$

while the remaining coefficients are given by the recurrence relation

$$\left[(\lambda' + 2k)^2 - m^2\right] C'_k + C'_{k-1} = \frac{C''_k}{2^m}, \quad k = 1, 2, 3, \cdots \tag{119}$$

where in virtue of Eq. (108), $C''_0 = \phi(0)$.

The Hertz vector component $\Pi_1^{(1)}$ is governed by Eq. (63), in which the coefficient of the second term in the equation to the first order of Δ is represented as

$$\frac{1}{\epsilon_1} \frac{\partial \epsilon_1}{\partial r} = -\frac{2\Delta}{a} \phi'\left(\frac{r_1}{u}\right) \tag{120}$$

The above expression follows from the definition of ε_1 in terms of the refractive index n given by Eq. (104). Thus, to a first order in Δ, Eq. (63) yields the equation

$$\nabla^2 \Pi_1^{(1)} - \frac{1}{c_1^2} \frac{\partial^2 \Pi_1^{(1)}}{\partial t^2} = \frac{2\Delta}{a} \phi'\left(\frac{r_1}{u}\right) \frac{\partial \Pi_3^{(1)}}{\partial z} = -\frac{2ik\Delta}{a} A \sum_{\nu=0}^{\infty} b_\nu \lambda_\nu J_m(r_1) J_1\left(\frac{\lambda_\nu r_1}{u}\right) \times e^{im\theta} e^{i(kz-\omega t)} \tag{121}$$

where the solution for $\Pi_3^{(1)}$ given by Eq. (109) is used, with the relation $J'_0(\cdot) = -J_1(\cdot)$. Hence, the particular solution of Eq. (121) is of the form

$$\Pi_1^{(1)} = f_1^{(g)}(r) e^{im\theta} e^{i(kz-\omega t)}. \quad 0 \le r \le a \tag{122}$$

which satisfies the differential equation

$$r_1^2 f_1^{(g)''}(r_1) + r_1 f_1^{(g)'}(r_1) + (r_1^2 - m^2) f_1^{(g)}(r_1) = -\frac{2iak\Delta}{u^2} A r_1^2 \sum_{\nu=0}^{\infty} b_\nu \lambda_\nu J_m(r_1) J_1\left(\frac{\lambda_\nu r_1}{u}\right) \tag{123}$$

The product of the Bessel functions on the right-hand side of Eq. (123) is now expressed as

$$J_m(r_1) J_1\left(\frac{\lambda_\nu r_1}{u}\right) = \frac{r_1^{m+1}}{2^{m+1}} \frac{\lambda_\nu}{u} \sum_{k=0}^{\infty} \frac{(-1)^k r_1^{2k} D_k^\nu}{4^k (k!)^2} \tag{124}$$

where

$$D_k^{(\nu)} = \begin{cases} \dfrac{k!}{(k+m)!} F\left(-k, -k-m, 2; \lambda_\nu^2/u^2\right), & \text{if} \quad \lambda_\nu < u \\ \dfrac{1}{m!(k+1)} \left(\dfrac{\lambda_\nu}{u}\right)^{2k} F\left(-k, -k-1, m+1; u^2/\lambda_\nu^2\right), & \text{if} \quad \lambda_\nu > u \end{cases} \tag{125}$$

([22], p. 960), in which the hypergeometric functions reduce to polynomials of degree k. Thus, the solution of Eq. (123) is of the form

$$f_1^{(g)}(r_1) = -B_2' \Delta A r_1^{\mu'} \left(D_0' + D_1' r_1^2 + D_2' r_1^4 + \cdots\right) \tag{126}$$

On substitution of the form (126) in Eq. (123) leads to the conclusion that

$$B_2' = \frac{2iak}{u^3} \tag{127}$$

$$\mu' = m + 3 \tag{128}$$

$$D_0' = \frac{D''_0}{2^{m+1}} \frac{1}{\mu'^2 - m^2} \tag{129}$$

and

$$\left[(\mu' + 2k)^2 - m^2\right] D_k' + D_{k-1}' = \frac{D''_k}{2^{m+1}}, \quad k = 1, 2, 3, \cdots \tag{130}$$

where

$$D''_k = \frac{(-1)^k}{4^k (k!)^2} \sum_{\nu=0}^{\infty} b_\nu \lambda_\nu^2 D_k^\nu \tag{131}$$

yielding the values of $D_0', D_1', D_2', \cdots$ in the solution of Eq. (126). Using forms (109) and (122) for the field components of the Hertz vector $\Pi_3^{(1)}$ and $\Pi_1^{(1)}$, respectively, in Eqs. (65)–(70), the expressions for the electromagnetic field components within the fiber core can be explicitly written down. The required components for the satisfaction of the components of the EM field are:

$$E_z^{(1)} = \frac{c_0^2}{n_1^2}\left(1 + \frac{2\Delta}{u^\alpha} r_1^\alpha\right) \frac{uA}{a^2} \Big[u\Big\{J_m(r_1) - B_1' \Delta r_1^{\lambda-2}\Big[C_0'(\lambda^2 - m^2) + C_1'\Big\{(\lambda+2)^2 - m^2\Big\} r_1^2$$
$$+ C_2'\Big\{(\lambda+4)^2 - m^2\Big\} r_1^4 + \cdots\Big]\Big\} + iakB_2'\Delta r_1^{\mu-1}\{D_0'(\mu+1) + D_1'(\mu+3) r_1^2 + D_2'(\mu+5) r_1^4 + \cdots\}\Big] \tag{132}$$

and

$$H_\theta^{(1)} = \frac{i\omega u A}{a\mu_1} \Big[J_m'(r_1) + B_1' \Delta r_1^{\lambda-1}\{\lambda C_0' + (\lambda+2) C_1' r_1^2 + (\lambda+4) C_2' r_1^4 + \cdots\}$$
$$- \frac{iak}{u} B_2' \Delta r_1^{\mu}\left(D_0' + D_1' r_1^2 + D_2' r_1^4 + \cdots\right)\Big] \tag{133}$$

In the extended medium of the cladding of the cable $r \geq a$, which is supposed to be of uniform index n_2 ($<n_1$), the Hertz vector $\Pi_3^{(2)}$ satisfies the wave equation (95). In as much as the electromagnetic field is totally axial in the medium with the radial component, $\Pi_3^{(2)} = \Pi_3^{(2)}\mathbf{e}_3$. For the diverging waves the solution for $\Pi_3^{(2)}$ is of the form given in Eq. (98).

As in the case of the core of the cable, the electromagnetic components $E_z^{(2)}$ and $H_\theta^{(2)}$ are given by the expressions

$$\begin{aligned} E_z^{(2)} &= -\frac{n_2^2}{c_0^2}\left[\frac{1}{r}\frac{\partial}{\partial r}\left(r\frac{\partial \Pi_3^{(2)}}{\partial r}\right) + \frac{1}{r^2}\frac{\partial^2 \Pi_3^{(2)}}{\partial \theta^2}\right] \\ &= -\frac{n_2^2}{c_0^2}\frac{v^2}{a^2} C K_m\left(\frac{v}{a}r\right) e^{im\theta} e^{i(kz-\omega t)} \end{aligned} \tag{134}$$

and

$$H_\theta^{(2)} = -\frac{1}{\mu_2}\frac{\partial}{\partial r}\left(\frac{\partial \Pi_3^{(2)}}{\partial t}\right) = \frac{i\omega v}{a\mu_2} C K_m'\left(\frac{v}{a}r\right) e^{im\theta} e^{i(kz-\omega t)} \tag{135}$$

5.1 The dispersion equation

As in Section 4.1, the boundary conditions $E_z^{(1)} = E_z^{(2)}$, and $H_\theta^{(1)} = H_\theta^{(2)}$ leads to the dispersion equation:

$$\begin{aligned} &(1+2\Delta)\left[\left\{J_m(u) - B_1'\,\Delta u^{\lambda-2}\left[C_0'(\lambda^2 - m^2) + C_1'\left\{(\lambda+2)^2 - m^2\right\}u^2 + C_2'\left\{(\lambda+4)^2 - m^2\right\}u^4 + \cdots\right]\right\}\right. \\ &\quad + iakB_2'\,\Delta u^{\mu-1}\left[D_0'(\mu+1) + D_1'(\mu+3)u^2 + D_2'(\mu+5)u^4 + \cdots\right]\Big] \times K_m'(v) \\ &\quad + u\left[J_m'(u) + B_1'\,\Delta u^{\lambda-1}\left\{\lambda C_0' + (\lambda+2)C_1'u^2 + (\lambda+4)C_2'u^4 + \cdots\right\}\right. \\ &\quad \left. - iakB_2'\,\Delta u^{\mu-1}\left(D_0' + D_1'u^2 + D_2'u^4 + \cdots\right)\right] \times v^2 K_m(v) = 0 \end{aligned} \tag{136}$$

Eq. (136) can be solved numerically for prescribed data. As before, let $n_1 = 1.5$, and $n_2 = 1.48515$. The equation is then expressed in terms of familiar numerical frequency f and the normalized propagation constant b defined in Eqs. (24), (25), and (26) with ak represented by Eq. (103). Thus, expressed in terms of f and b, the zeros of the left-hand side of Eq. (136) can be found as stated in Section 1. The dispersion curves for specified form of the profile ϕ can then be obtained for different modes m.

As a specific application, consider a profile of the refractive index similar to that of the Nonzero Dispersive Shifted Fiber (NZDSF) ([23], p. 32) represented by the function

$$\phi\left(\frac{r}{a}\right) = \begin{cases} = 1, & 0 \leq r/a \leq 1/4 \\ = -1/3, & 1/4 < r/a \leq 1/2 \\ = 1/6, & 1/2 < r/a \leq 3/4 \\ = -1/12, & 3/4 < r/a \leq 1 \end{cases} \tag{137}$$

Performing detailed computation of the left-hand side of Eq. (136) for the relevant fundamental mode $m = 0$, it is found that the expresssion is nonzero for b ranging from 0 to 1 and f ranging from 0 to 10, indicating that there is absolutely no chromatic wave guide dispersion of the propagating wave. This means that a pulse of wave would pass without distortion of shape through core of the cable in the present case for the data considered.

6. Conclusion

The theory of light propagating through an optical fiber coated with a thin layer of slightly less refractive index for the purpose of of total internal reflection is presented in this chapter. The development of the theory is based on the Maxwell equations whose solution is represented by the Hertz vector Π ([8], p. 28), which in this particular case is oriented along the axis of the core of the fiber. For a homogeneous core step-index fiber, the vector function Π satisfies the wave equation of light propagation (Eq. (12)), while for a graded-index inhomogeneous core, a slight modification of representation yields a modified form of wave equation (Eq. (61)). The electromagnetic field intensities E and H are derivable from Π. Modal analysis of the relevant wave equation using appropriate boundary conditions at the interface between the core and the cladding yields the desired chromatic dispersion equation. At first, the step-index fiber is analyzed by this procedure, to obtain the equation in a nondimensional form of frequency and phase velocity. It is then numerically solved for the first two modeled $m = 0$ and $m = 1$ for representative values of the refractive indices to find that while the waves travel in groups, their velocities vanish for certain frequency ranges. From this observation, one concludes that during transmission of data in the dispersive modes, it may result in bursts in transmission with the possibility of disruption in the transmission itself. Next to this study, the macrobending effects for small bends are investigated, assuming that the curvature is small, and it is found that in general loss of power of transmission takes place due to the bends and so such bends should better be avoided as far as possible in cable laying. Finally, the theory of light wave propagation in inhomogeneous graded-index cladded fiber is carried out in order to understand its chromatic dispersion characteristics. The corresponding modified wave equation for such media is treated for the much discussed parabolic profile of some degree α of the refractive index ([3], p. 54). The dispersion equation is developed for the propagation through the core of this type and the numerical treatment of the prototype case for the first two modes $m = 0$ and $m = 1$ reveals that the dispersion phenomenon still takes place but on a very reduced scale. Finally, the method of solution of the parabolic case is generalized to the case of general form of the refractive index of the core material. The method of solution is applied to the case of a profile similar to that of an NZDSF fiber to find that no dispersion takes place in the fundamental mode of propagation. To conclude, reduced dispersion of light propagation is a desired property for data transmission through a fiber-optic cable. In applications however a high number of modes are employed for increased bandwidth. These features can be extracted from the general dispersion equations developed in this chapter.

Acknowledgements

The author is thankful to the SN Bose National Center for Basic Sciences for supporting this research, as also to the editor for valuable advice for better presentation of the chapter.

Disclosures

There are no conflicts of interest involved in the presented chapter.

Data availability

No data were generated or analyzed by Artificial Intelligence (AI) or otherwise in the presented chapter.

Author details

Sujit K. Bose
SN Bose National Center for Basic Sciences, Kolkata, India

*Address all correspondence to: sujitkbose1@gmail.com

References

[1] Marcuse D. Light Transmission Optics. New York: Van Nostrand Reinhold; 1982

[2] Snyder AW, Love JD. Optical Wave Guide Theory. London: Chapman and Hall; 1983

[3] Keiser G. Optical Fiber Communication. New York: McGraw-Hill; 2011

[4] Saleh BEA, Teich MC. Fundamentals of Photonics. Hoboken: Wiley; 2019

[5] Agrawal GP. Fiber-Optic Communication Systems. Hoboken: John Wiley; 2010

[6] Born M, Wolf E. Principles of Optics. Cambridge: Cambridge University Press; 2005

[7] Bose SK. On the transmission of data packets through fiber-optic cables of uniform index. Journal of Optical Communications. 2021;**44**:s1647-s1651

[8] Stratton JA. Electromagnetic Theory. New Jersey: John Wiley; 2007

[9] Willinger W, Paxson V, Taqqu MS. Self-similarity and heavy tails: Structural modelling of traffic. In: Structural Techniques and Applications. Boston: Birkhauser; 1998

[10] Thompson K, Miller GJ, Wilder R. Wide-area internet traffic patterns and characteristics. IEEE Networks. 1997;**11**: 10-23. DOI: 10.1109/65.642356

[11] Kim M-S, Won YJ. Characteristic analysis of internet traffic from the perspective of flows. Computer Communications. 2005;**XX**:1-14

[12] Bose SK. Theoretical macrobending loss in single-mode transmission through a uniform index fiber-optic cable. Optimal Control. 2022;**1**:2522-2532. DOI: 10.1364/OPTCON473789

[13] Bose SK. Multimode dispersion of light wave propagation in graded-index cladded fiber-optic cable. Universal Journal of Applied Mathematics. 2024; **12**:95-108. (to appear)

[14] Miller DAB. Wave modes, communications, and optics: A tutorial. Advances in Optics and Photonics. 2019; **11**:679-825. DOI: 10.1364/AOP.11.000679

[15] Cao H, Cizmar T, Turtaev S, Tyc T, Rotter S. Controlling light propagation in multimode fibers for imaging, spectroscopy and beyond. Advances in Optics and Photonics. 2023;**15**:524-612. DOI: 10.1364/AOP.484298

[16] Abramowitz M, Stegun IA. Handbook of Mathematical Functions. New York: Dover Publications; 1972

[17] Bhadra S, Ghatak A. Wave Optics and Photonic Devices. Boca Raton: CRC Press; 2013

[18] Leland WE, Taqqu MS, Willinger DK. On the self-similar nature of internet traffic. Computer Communication Review. 1995;**25**: 202-213

[19] Nayfeh AH. Perturbation Methods. New York: John Wiley; 1973

[20] Mohsen A. Electromagnetic field representation in inhomogeneous anisotropic media. Applied Physics. 1973;**3**:123-128

[21] Watson GN. A Treatise on the Theory of Bessel Functions. Cambridge: Cambridge University Press; 1952

[22] Gradshteyn IS, Ryzhik IM. Table of Integrals, Series and Products. Orlando: Academic Press; 1980

[23] Cvijetic M. Optical Transmission. Norwood: Artech House; 2004

Chapter 2

Higher-Order Raman Pure-Quartic Wave Trains in Optical Fibers

Conrad Bertrand Tabi, Karabo Kefilwe Ndebele, Hippolyte Tagwo, Gaston Camus Latchio Tiofack and Timoléon Crépin Kofane

Abstract

This chapter addresses the effects of higher-order Raman scattering on modulational instability in optical fibers with pure-quartic dispersion. Low- and high-input power regimes are affected by self-frequency shift (SFS) effects and, due to the presence of the quintic nonlinearity and quartic dispersion, this leads to extended zones of instability in the gain spectrum. Here, we demonstrate *via* numerical simulations that our biharmonic nonlinear Schrödinger (NLS) equation exhibits ultrashort pulse trains, including rogue wave (RW)-like patterns, as well as trains of Akhmediev breathers (ABs). The high-input power can also lead to a tunable strong Raman-induced frequency shift, allowing energy transfer between modes during signal propagation. This work paves the way for a potential experimental characterization of higher-order non-Kerr nonlinearities in pure-quartic optical media.

Keywords: optical fibers, pure-quartic solitons, gain of instability, higher-order Raman effects, modulational instability

1. Introduction

The recent experimental characterization of pure-quartic solitons (PQSs) has generated increased interest in their potential applications in nonlinear optics and their various branches [1, 2]. PQSs have unique properties due to their dominant fourth-order dispersion (FOD), which sets them apart from conventional solitons. They have the ability to maintain their temporal and spectral shapes, making them a viable option for engineering high-power ultrashort pulses [3–5].

Over the years, it has been confirmed that the balance between Kerr nonlinearity and negative second-order dispersion (SOD) creates stable temporal solitons [6]. On the other hand, PQSs are formed when there is a balance between negative FOD and positive Kerr nonlinearity without the influence of other forms of nonlinearities [1, 7]. However, recent advances in the quest for PQS pulses of less than one picosecond have motivated the inclusion of other forms of nonlinearities, such as quintic nonlinearity [8], intrapulse Raman scattering [9, 10], and nonlocality [11, 12], to name a few. In addition, when the intensity of the incident light field increases, it can cause non-Kerr nonlinear effects [13, 14]. These effects can have a significant impact on the formation and stability of PQSs, thereby resulting in soliton-like propagation of femtosecond optical pulses [1, 3, 15]. This has inspired this contribution, which examines cubic-quintic Raman effects within the modulational instability (MI) process. The aim is to understand how these effects relate to the emergence of pure-quartic solitonic patterns in optical fibers. To remind, MI is a common nonlinear wave phenomenon observed in various fields such as hydrodynamics [16–19], optics [20–25], plasmas [26–29], cold atom condensates [30–35], topological insulators [36], and biophysics [37–42]. It is characterized by the rapid growth of slow periodic perturbations over a background wave. In recent works dedicated to PQSs, effects related to quintic nonlinearity [8], weak and strong nonlocalities [12], cubic Raman scattering [10], and birefringence [43] have been explored. Particularly, it was shown that the FOD could cancel the frequency redshift due to the Raman effect on the propagation of ultrashort rogue-wave (RW) trains, while the quintic nonlinearity contributed to extending the occurrence of MI to positive values of the quartic-dispersion coefficient.

In this chapter, we use linear stability analysis (LSA) of continuous wave (CW) to differentiate between low-input power (LIP) and high-input power (HIP) regimes for pure-quartic pattern formation. Our results show that a suitable balance between the quintic Raman effect and FOD supports low-frequency bandwidths in the LIP mode, while the HIP regime accommodates extended and high-frequency bandwidths. Numerical simulations on the higher-order biharmonic nonlinear Schrödinger (NLS) equation corroborate the analytical predictions.

2. Model and linear stability analysis

To proceed, we consider the ultrashort pulse propagating in an optical fiber to be modeled by the following higher-order biharmonic NLS equation:

$$\begin{aligned} iE_z + \frac{\beta_4}{24}E_{tttt} + \gamma_1|E|^2E - i\alpha_1\left(|E|^2E\right)_t \\ -i\alpha_2E\left(|E|^2\right)_t + \gamma_2|E|^4E - i\alpha_3\left(|E|^4E\right)_t \\ -i\alpha_4E\left(|E|^4\right)_t = 0, \end{aligned} \tag{1}$$

where $E(z,t)$ represents the complex envelope of the electric field, with z and t being the propagation distance and time, respectively. β_4 is the FOD coefficient. $\gamma_1 = 4.072\ \mathrm{W}^{-1}\ \mathrm{km}^{-1}$ and γ_2 are, respectively, the cubic and quintic nonlinearities, respectively, while $\alpha_1 = -0.0247\ \mathrm{kW}^{-1}/[(2\pi)\mathrm{mTHz}]$ and $\alpha_3 = -0.0247\ \mathrm{kW}^{-2}/[(2\pi)\mathrm{mTHz}]$ are the cubic and quintic self-steepening coefficients, respectively. $\alpha_2 = 0.03705\ \mathrm{kW}^{-1}\mathrm{psm}^{-1}$ and α_4 are, respectively, the cubic and quintic intrapulse Raman scattering terms, respectively. To remind, the simple biharmonic NLS equation can be retrieved by considering $\alpha_1 = \alpha_2 = \alpha_3 = \alpha_4 = \gamma_2 = 0$ [1, 7, 44]. In this case, the corresponding experimental setup can be realized *via* a recirculating fiber loop, in which linear losses are compensated *via* Raman amplification [44]. Also, when $\alpha_3 = \alpha_4 = \gamma_2 = 0$, one recovers the model proposed in Refs. [9, 10] that was used to characterize the frequency redshift due to the cubic Raman scattering.

We are interested in CW radiations for which the amplitude $E(z,t)$ remains time-independent at the input end of the fiber at $z = 0$. We, therefore, proceed by assuming the same hypothesis when the signal propagates inside the fiber, which leads to the steady-state solution $E(z,t) = \sqrt{P_0}e^{i\phi_{NL}}$, where P_0 is the input power and $\phi_{NL} = (\gamma_1 P_0 + \gamma_2 P_0^2)z$ is the nonlinear phase shift induced by self-phase modulation (SPM) and non-Kerr quintic nonlinearity. The stability of the steady-state solution can be regarded by introducing small perturbations so that $E(z,t) = \left[\sqrt{P_0} + \delta E(z,t)\right]e^{i\phi_{NL}z}$, where the perturbative complex field $\delta E(z,t)$ is such that $|\delta E(z,t)| \ll \sqrt{P_0}$. The perturbation is such that $\delta E(z,t) = e_1 e^{i(Kz-\Omega t)} + e_2 e^{-i(Kz-\Omega t)}$, where e_1 and e_2 are real constant coefficients. Ω and K are, respectively, a complex frequency and wavenumber of the perturbation. The standard procedure of LSA leads to the dispersion relation

$$K = P_0 a_1 \Omega + \frac{|\Omega|}{24}\sqrt{576P_0^2 a_2^2 + 48 n_1 \beta_4 \Omega^2 + \beta_4^2 \Omega^6}, \tag{2}$$

where $a_1 = 2\alpha_1 + \alpha_2 + P_0(3\alpha_3 + 2\alpha_4)$, $a_2 = \alpha_1 + \alpha_2 + 2P_0(\alpha_3 + \alpha_4)$, and $n_1 = \gamma_1 P_0 + 2\gamma_2 P_0^2$. MI will, therefore, occur in the system only if the wavenumber possesses a nonzero imaginary part, leading to an exponential growth of the perturbed amplitude measured by the MI gain $\zeta(\Omega) = 2\mathrm{Im}[K(\Omega)]$, that is,

$$\zeta(\Omega) = \frac{|\Omega|}{12}\sqrt{576P_0^2 a_2^2 + 48 n_1 \beta_4 \Omega^2 + \beta_4^2 \Omega^6}. \tag{3}$$

The above condition depends strictly on the expression under the square root so that $n_1\beta_4 > 0$. Therefore, a first case imposes $\beta_4 < 0$ and $\gamma_2 < 0$, while a second combination suggests $\beta_4 > 0$ and $\gamma_2 < 0$, given that $\gamma_1 > 0$ and $P_0 > 0$. The MI gains corresponding to the two cases are shown in **Figure 1(a)** and **(b)**, respectively. One can therefore notice a restriction on the input power, where LIP regime occurs for $(\beta_4 < 0, \gamma_2 < 0)$, while the HIP regime takes place for $(\beta_4 > 0, \gamma_2 < 0)$. The combined role of the quintic nonlinearity and FOD on the occurrence of the MI gain was discussed recently, and it was proposed that such a system could support MI for $\beta_4 > 0$ for negative γ_2 [8], which corresponds to our second case. However, we suspect the higher-order Raman (cubic-quintic) terms to be responsible for the first case. For precision, the MI gain, in a more generalized model that includes higher-order even and odd dispersions, was proposed [21]. It was then found that, in more generalized models of optical fibers, odd dispersions do not contribute to the MI gain. Remarkably, for PQS models, only the FOD contributes to the process and impacts the distribution of the MI gain. For the case at hand, the MI sidebands are related to the following optimum modulation frequency (OMF), obtained for $d\zeta/d\Omega = 0$, that is,

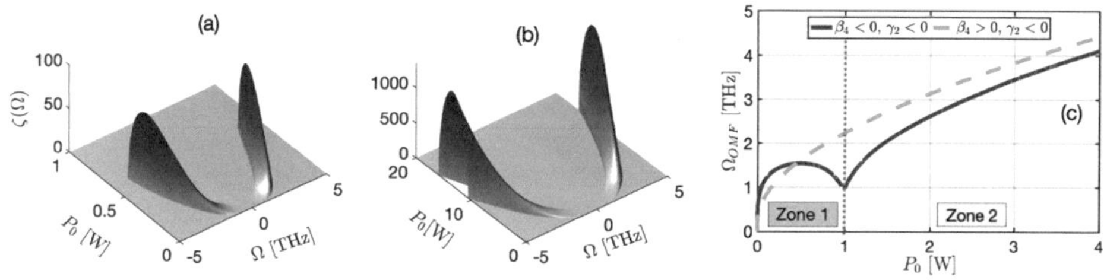

Figure 1.
(a)-(b) MI gain for the LIP and HIP regimes, respectively. (c) OMF for the LIP (blue line) and HIP (dashed yellow line) regimes, with $|\beta_4| = 2.2\ ps^4 km^{-1}$, $|\gamma_2| = 10^{-3}\ W^{-2}\ km^{-1}$, $|\alpha_4| = 0.030875\ W^{-1} ps km^{-1}$, and the other parameter values being given in the text.

$$\Omega^6 + \frac{24n_1}{\beta_4}\Omega^2 + \frac{144a_2^2P_0^2}{\beta_4^2} = 0. \qquad (4)$$

By solving the above, one obtains the OMF Ω_{OMF}, which is also plotted in **Figure 1 (c)** versus the input power P_0, for the two reported cases. For the case $(\beta_4 < 0, \gamma_2 < 0)$, the presence of a critical value of the input power $P_{0,cr}$ is clearly indicated and divides two zones: the LIP zone (**Zone 1**) and the HIP zone (**Zone 2**). The case $(\beta_4 > 0, \gamma_2 < 0)$ shows an increase in the OMF conventionally known to happen in the absence of singularities. This goes beyond the predictions by Cavalcanti et al. [45], who suggested that in the absence of SOD, the FOD system emerges due to a negative β_4 and the OMF saturates to a given value by $\Omega_{OMF} = \sqrt{48\gamma_1 P_0/|\beta_4|}$.

Following the same procedure, the MI gain for the two regimes is plotted in **Figures 2** and **3**, versus the modulation frequency Ω and the quintic SFS coefficient α_4. In **Figure 2**, where $\beta_4 < 0$ and $\gamma_2 < 0$, the panels correspond to increasing low values of the input power, which lead to a restricted frequency bandwidth and low, but increasing, MI gain. Noticeably, the frequency bandgap between the symmetric breast of instability grows with P_0, while instability decreasingly gets restricted to small intervals of α_4, with the maximum at $\alpha_4 = 0$. The same spectrum of behaviors is delivered in **Figure 3**, under the HIP regime, with the panels corresponding, respectively, to the values 2, 5, and 10 W of P_0. The lobes of instability appear in zones of high modulation frequency, and the MI gain is very sensitive to increasing P_0. Calculations have also been made for the MI gain to be represented against the modulation frequency Ω and α_3 as depicted in **Figure 4**, where panel (a) corresponds to the LIP regime and panel (b) to the HIP mode. In this context, the difference between the modes is obvious, where the MI gain supports the appearance of satellite sidebands of high intensity at extended modulation frequencies. Similarly, different combinations of other parameters can be exploited. However, the most important result to remember is the possibility of positive values of β_4 leading to MI. When parameter values are chosen from such zones, the CW will break up into solitary waves and be said to be unstable under modulation, while it will remain stable in zones where the MI gain does not occur.

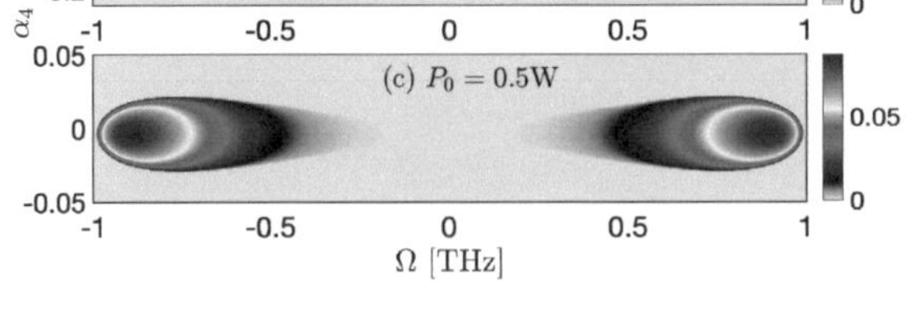

Figure 2.
MI gain for the LIP regime versus the modulation frequency Ω and the quintic SFS α_4, with increasing values of P_0, using the parameter values: $\beta_4 = -2.2\ ps^4 km^{-1}$, $\gamma_2 = -10^{-3}\ W^{-2}\ km^{-1}$, and the other parameters being given in the text.

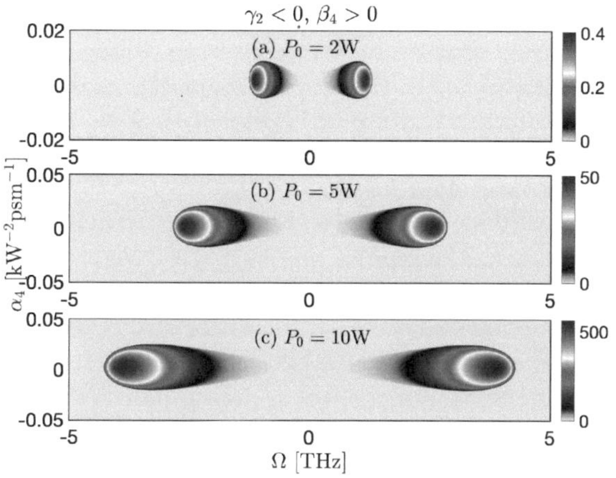

Figure 3.
MI gain for the HIP regime versus the modulation frequency Ω *and the quintic SFS* α_4*, with increasing values of* P_0*, using the parameter values:* $\beta_4 = 2.2\ ps^4km^{-1}$*,* $\gamma_2 = -10^{-3}\ W^{-2}\ km^{-1}$*, and the other parameters being given in the text.*

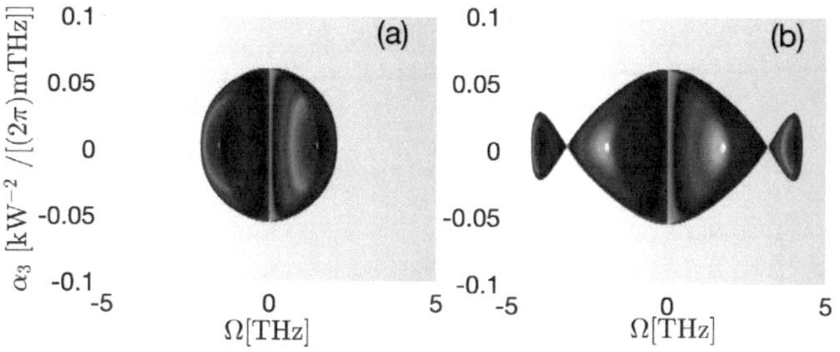

Figure 4.
MI gain for the HIP regime versus the modulation frequency Ω *and the quintic self-steepening coefficient* α_3*, where panel (a) corresponds to the LIP regime and panel (b) refers to the HIP regime, with* $|\beta_4| = 2.2\ ps^4km^{-1}$*, and* $\gamma_2 = -10^{-3}\ W^{-2}\ km^{-1}$*. For (a)* $P_0 = 0.08$ *W and (b) is recorded for* $P_0 = 5$ *W, with the other parameters being given in the text.*

3. Numerical simulations

To verify the MI analysis, we numerically solve the generic Eq. (1) using the CW input signal $E(z=0) = \sqrt{P_0} + a_m \sin(\Omega_m t)$, where P_0 takes values from the appropriate input power regime, and $a_m = 10^{-3}$ is the modulation amplitude. Ω_m is the modulation frequency whose values should fall inside the instability zones of **Figures 1–3**. The split-step Fourier method is used, and attention is given to the combined effects between the FOD, the quintic nonlinearity, and the quintic SFS. In all simulations, the signal is propagated over a distance $z = 1000L_{FOD}$, with $L_{FOD} = T_0^4/|\beta_4|$ and T_0 being the half-width at the $1/e-$ intensity point of the signal maximum [1].

For the HIP regime, we fix $P_0 = 3$ W, $\Omega_m = 3.5$ THz, and $\gamma_1 = 4.072\ \text{W}^{-1}\ \text{km}^{-1}$. In general, MI is supported by the appearance of many subpulses in the temporal domain, which clearly confirms predictions from the LSA, as per **Figure 5**. Exclusively, by suitably balancing γ_2 and β_4, fair competition between α_2 and α_4 can allow for the control of the directional propagation of the modulated CW signal. **Figure 5** $(aj)_{j=1,2}$ and $(bj)_{j=1,2}$ correspond, respectively, to $\alpha_4 = 0.030875$

and $\alpha_4 = -0.030875\ W^{-1}pskm^{-1}$. This is solidly supported by the spectral powers that show energy exchange characteristics between the emerging sidebands of RW trains (see **Figure 5(a2)** and **(b2)**). Correspondingly, the redshift changes direction, which justifies the background shifting of **Figure 6(a)**. Also, the maximum intensity experiences exponential growth and corroborates the shifting reported in **Figure 6(a)** over the propagation distance (see **Figure 6(b)**).

For the LIP regime, numerical experiments are carried out by considering $P_0 = 0.05$ W and $\Omega_m = 0.75$ THz. The regime refers to $\gamma_2 < 0$ and $\beta_4 > 0$, while the quintic SFS α_4 can be positive or negative. The corresponding results are displayed in **Figure** 7, where the left panels show MI patterns for $\alpha_4 > 0$ and the right panels correspond to $\alpha_4 < 0$. Like in the previous case, the CW disintegrates into trains of ultrashort pulses, with an obvious signature of breathing RWs with reduced frequency compared to the case of **Figure 5**. This gives rise to extended ABs, while the whole process is favored by energy exchange between the modes as testified by the spectral behaviors of **Figure** 7**(a2)**. Indeed, as already noticed, the propagation direction

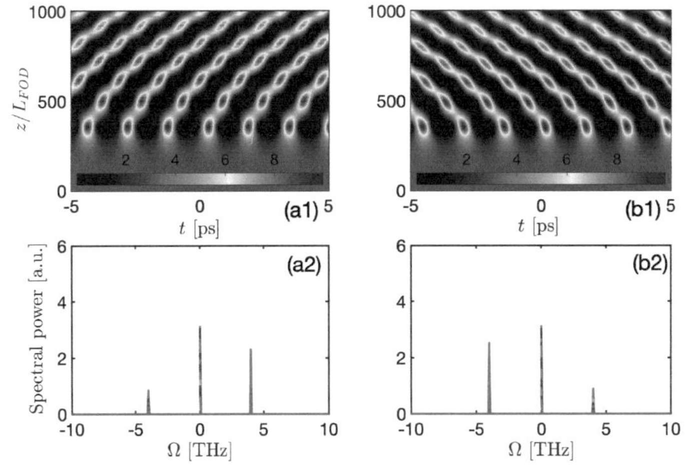

Figure 5.
Effect of the quintic SFS sign on MI under HIP regime, where the upper line shows the spatiotemporal evolution and the lower line displays the spectral evolution of unstable modes, with (a1)-(a2) $\alpha_4 = 0.030875\ W^{-1}pskm^{-1}$, and (b1)-(b2) $\alpha_4 = -0.030875\ W^{-1}pskm^{-1}$. The other parameters are: $P_0 = 3$ W, $\Omega_m = 3.5$ THz, $\beta_4 = 2.2\ ps^4km^{-1}$, and $\gamma_2 = -10^{-3}\ W^{-2}\ km^{-1}$.

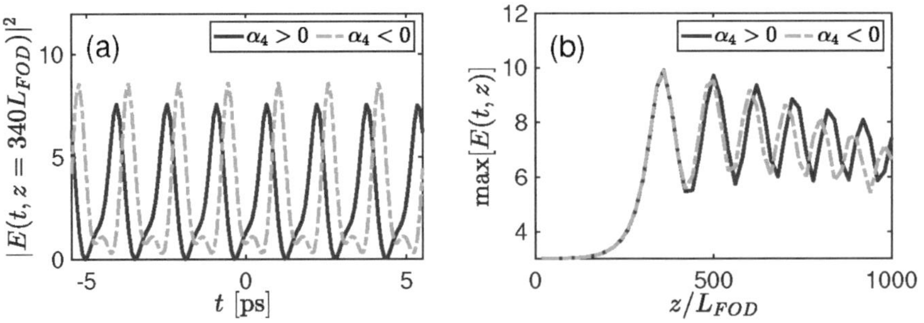

Figure 6.
*Temporal cross section of the wave patterns in **Figure** 5(a1) and (b1) at distance $z = 340L_{FOD}$. The maximum amplitudes of the patterns are compared in panel (b), with values of parameters being the same as in **Figure** 5.*

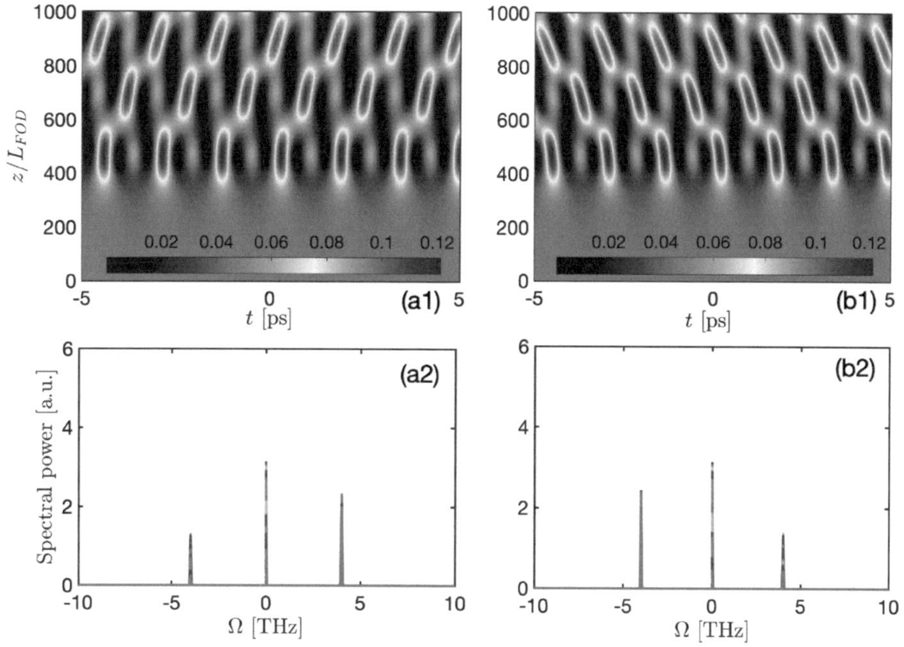

Figure 7.
Effect of the quintic SFS sign on MI under LIP regime, where the upper line shows the spatiotemporal evolution and the lower line displays the spectral evolution of unstable modes, with (a1)-(a2) $\alpha_4 = 0.030875\ W^{-1}pskm^{-1}$, and (b1)-(b2) $\alpha_4 = -0.030875\ W^{-1}pskm^{-1}$. The other parameters are: $P_0 = 0.05\ W$, $\Omega_m = 0.75\ THz$, $\beta_4 = -2.2\ ps^4km^{-1}$, and $\gamma_2 = -10^{-3}\ W^{-2}\ km^{-1}$.

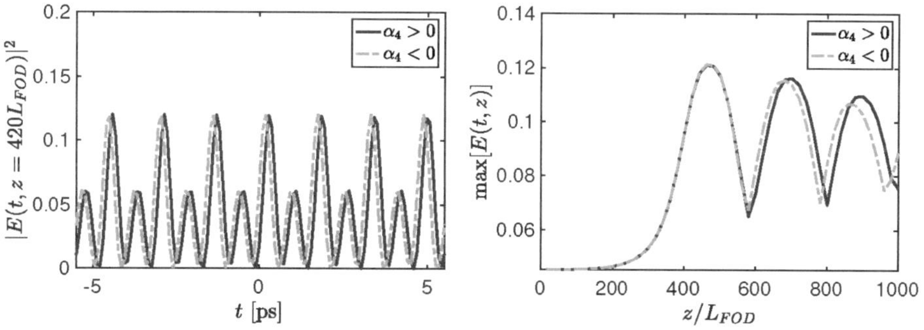

Figure 8.
*Temporal cross section of the wave patterns in **Figure** 7(a1) and (b1) at distance $z = 420L_{FOD}$. The maximum amplitudes of the patterns are compared in panel (b), with values of parameters being the same as in **Figure** 7.*

changes when $\alpha_4 < 0$, which shows some signature of frequency redshift in **Figure 7 (b1)**, along with the spectral behaviors of **Figure 7(b2)**. The emergence of trains of RWs is corroborated by the results of **Figure 8(a)**, where the temporal cross sections of **Figure 7(a1)** and **(b1)** are compared. Beyond the subsequent wave shift between the two cases of positive and negative α_4, **Figure 8(b)** displays an exponential growth of the perturbation, but only three trains of waves appear, while their shifting process takes progressively place as z increases.

Mixed modes can emerge in unstable conditions with high MI gain, causing multifrequency generation [46]. This is facilitated by inherent parameters of the optical fiber system and may result in higher Raman orders. For example, increasing

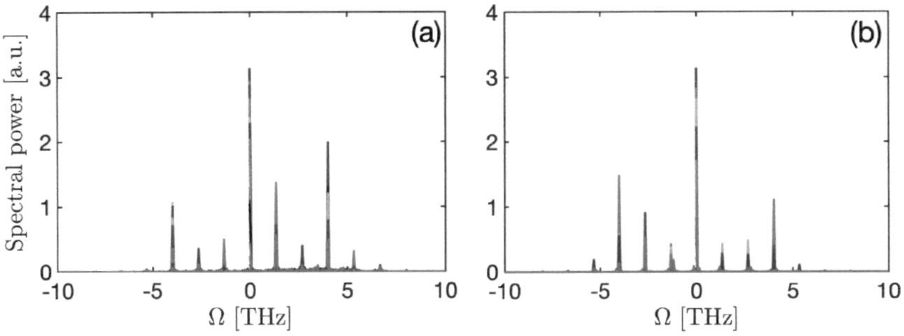

Figure 9.
The panels show the spectral manifestation of MI under the effect of the quartic dispersion β_4 for $\alpha_4 = 0.0825$ $W^{-1}pskm^{-1}$: (a) $\beta_4 = -2.2\ ps^4km^{-1}$ and (b) $\beta_4 = -2.8\ ps^4km^{-1}$.

$|\beta_4|$ in **Figure 9** significantly alters the effect of SFS, resulting in frequency-comb sidebands. In other words, new optical frequencies are generated, which is beneficial for maintaining signal integrity over long distances by selecting longer wavelengths. Pure-quartic dispersive materials with higher-order Raman scattering effects are necessary to generate alternative input power regimes. This can enable experimental control over the speed and direction of optical signals [47].

4. Conclusion

To sum up, we explored MI in optical fibers *via* a higher-order biharmonic NLS equation containing higher-order Raman scattering terms. Using the LSA, we detected LIP and HIP regimes, with the latter producing RW subpulses and the former producing ultrashort pulses. In addition, trains of ABs have been generated. Such results were confirmed in the spectral domain, confirming the possibility of balancing the cubic and quintic SFSs to monitor the direction of the optical beam for both positive and negative quartic dispersions. Our results open new routes to understanding the effect of higher-order Raman scattering terms on the generation and propagation of PQSs and their applications to silica photons crystals and photonic crystal waveguides.

Author details

Conrad Bertrand Tabi[1*†], Karabo Kefilwe Ndebele[1†], Hippolyte Tagwo[2†], Gaston Camus Latchio Tiofack[3†] and Timoléon Crépin Kofane[1,4†]

1 Department of Physics and Astronomy, Botswana International University of Science and Technology, Botswana

2 University Institute of Wood Technology, Mbalmayo, Cameroon

3 Faculty of Sciences, University of Maroua, Maroua, Cameroon

4 Department of Physics, University of Yaoundé I, Yaoundé, Cameroon

*Address all correspondence to: tabic@biust.ac.bw

† These authors contributed equally.

References

[1] Blanco-Redondo A, de Sterke CM, Sipe J, et al. Pure-quartic solitons. Nature Communications. 2016;7:10427

[2] Runge AFJ, Hudson DD, Tam KKK, et al. The pure-quartic soliton laser. Nature Photonics. 2020;**14**:492

[3] Taheri H, Matsko AB. Quartic dissipative solitons in optical Kerr cavities. Optics Letters. 2019;**44**:3086

[4] Melchert O, Yulin A, Demircan A. Dynamics of localized dissipative structures in a generalized Lugiato–Lefever model with negative quartic group-velocity dispersion. Optics Letters. 2020;**45**:2764

[5] Zhao K, Gao C, Xiao X, Yang C. Vector quartic solitons in birefringent fibers. Optics Letters. 2021;**46**:761

[6] Agrawal G. Applications of Nonlinear Fiber Optics. Massachusetts, United States of America: Academic Press, Elsevier Science; 2020

[7] Tam KK, Alexander TJ, Blanco-Redondo A, Martijn de Sterke C. Stationary and dynamical properties of pure-quartic solitons. Optics Letters. 2019;**44**:3306

[8] Soltani M, Triki H, Azzouzi F, et al. Pure–quartic optical solitons and modulational instability analysis with cubic–quintic nonlinearity. Chaos, Solitons & Fractals. 2023;**169**:113212

[9] Wang Z, Luo C, Wang Y, et al. Raman-induced frequency shift of pure quartic solitons in optical fiber with quartic dispersion. Optics Letters. 2022;**47**:3800

[10] Tiofack CGL, Tabi CB, Tagwo H, Kofané TC. Pure quartic wave modulation in optical fiber with the presence of self-steepening and intrapulse Raman scattering response. Physics Letters A. 2023;**480**:128982

[11] Triki H, Pan A, Zhou Q. Pure-quartic solitons in presence of weak nonlocality. Physics Letters A. 2023;**459**:128608

[12] Tiofack CGL, Tabi CB, Tagwo H, Kofané TC. Nonlocal cubic and quintic nonlinear wave patterns in pure-quartic media. Journal of Optics. 2023;**25**:054001

[13] Mitschke FM, Mollenauer LF. Discovery of the soliton self-frequency shift. Optics Letters. 1986;**11**:659

[14] Gordon JP. Theory of the soliton self-frequency shift. Optics Letters. 1986;**11**:662

[15] Runge AFJ, Qiang YL, Alexander TJ, et al. Infinite hierarchy of solitons: Interaction of Kerr nonlinearity with even orders of dispersion. Physical Review Research. 2021;**3**:013166

[16] Benjamin TB, Feir JE. The disintegration of wave trains on deep water part 1. Theory. Journal of Fluid Mechanics. 1967;**27**:417

[17] Bansi Kamdem CD, Ndjawa Yomi PA, Tabi CB, Mohamadou A. Modulated blood waves in the coupled complex Ginzburg–Landau equations of Jeffrey fluids in arteries. European Physical Journal Plus. 2023;**138**:176

[18] Edouma Biloa BP, Tabi CB, Ekobena Fouda HP, Kofané TC. Nonlinear dissipative wave trains in a system of self-propelled particles. Physica Scripta. 2023;**98**:115230

[19] Bansi CD, Tabi CB, Mohamadou A. Dissipative Mayer's waves in fluid-filled viscoelastic tubes. Chaos, Solitons & Fractals. 2018;**109**:170

[20] Tai K. Observation of modulational instability in optical fibers. Physical Review Letters. 1986;**56**:135

[21] Ndebele KK, Tabi CB, Tiofack CGL, Kofané TC. Higher-order dispersion and nonlinear effects of optical fibers under septic self-steepening and self-frequency shift. Physical Review E. 2021;**104**: 044208

[22] Abemgnigni Njifon M, Tabi CB, Kofané TC. Few-cycle optical pulses in negative index materials with dispersive permittivity and permeability. Journal of the Optical Society of America B: Optical Physics. 2020;**37**:A331

[23] Megne Tiam L, Tabi CB, Kofané TC. Modulation instability in nonlinear metamaterials modeled by a cubic-quintic complex Ginzburg-Landau equation beyond the slowly varying envelope approximation. Physical Review E. 2020;**102**:042207

[24] Tiam Megne L, Tabi CB, Ambassa Otsobo JA, Muiva CM, Kofané TC. Propagation of dissipative simple vortex-, necklace- and azimuthon-shaped beams in Kerr and non-Kerr negative-refractive-index materials beyond the slowly varying envelope approximation. Nonlinear Dynamics. 2023;**111**:20289

[25] Zanga D, Fewo SI, Tabi CB, Kofané TC. Generation of dissipative solitons in a doped optical fiber modeled by the higher-order dispersive cubic-quintic-septic complex Ginzburg-Landau equation. Physical Review A. 2022;**105**:023502

[26] Panguetna CS, Tabi CB, Kofané TC. Two-dimensional modulated ion-acoustic excitations in electronegative plasmas. Physics of Plasmas. 2017;**24**: 092114

[27] Zabolotnykh AA. Nonlinear Schrödinger equation for a two-dimensional plasma: Solitons, breathers, and plane wave stability. Physical Review B. 2023;**108**:115424

[28] Tabi CB, Panguetna CS, Kofané TC. Modulational instability in nonlinear saturable media with competing nonlocal nonlinearity. Physica B. 2018; **5C**:370

[29] Panguetna CS, Tabi CB, Kofané TC. Electronegative nonlinear oscillating modes in plasmas. Communications in Nonlinear Science and Numerical Simulation. 2018;**55**:326

[30] Madimabe EB, Tabi CB, Tiofack CGL, Kofané TC. Modulational instability in vector exciton-polariton condensates with photonic spin-orbit coupling. Physical Review B. 2023;**107**: 184502

[31] Tabi CB, Veni S, Wamba E, Kofané TC. Modulational instability and droplet formation in Bose-Bose mixtures with Lee-Huang-Yang correction and polaron-like impurity. Physics Letters A. 2023;**485**:129087

[32] Otlaadisa P, Tabi CB, Kofané TC. Modulation instability in helicoidal spin-orbit coupled open Bose-Einstein condensates. Physical Review E. 2021; **103**:052206

[33] Tabi CB, Veni S, Kofané TC. Generation of matter waves in Bose-Bose mixtures with helicoidal spin-orbit coupling. Physical Review A. 2021;**104**: 033325

[34] Tabi CB, Otlaadisa P, Kofané TC. Modulation instability of two-dimensional

Bose-Einstein condensates with helicoidal and a mixture of Rashba-Dresselhaus spin-orbit couplings. Physics Letters A. 2022; **449**:128334

[35] Tabi CB, Wamba E, Nare E, Kofané TC. Interplay between spin-orbit couplings and residual interatomic interactions in the modulational instability of two-component Bose-Einstein condensates. Physical Review E. 2023;**107**:044206

[36] Leykam D. Probing band topology using modulational instability. Physical Review Letters. 2021;**126**:048901

[37] Tabi CB, Etémé AS, Kofané TC. Unstable cardiac multi-spiral waves in a FitzHugh–Nagumo soliton model under magnetic flow effect. Nonlinear Dynamics. 2020;**100**:3799

[38] Zaoro NR, Tabi CB, Etémé AS, Kofané TC. Unstable cAMP wave patterns during aggregation of Dictyostelium discoideum cells. Physics Letters A. 2020;**384**:126133

[39] Okaly JB, Mvogo A, Tabi CB, Ekobena Fouda HP, Kofané TC. Base pair opening in a damped helicoidal Joyeux-Buyukdagli model of DNA in an external force field. Physical Review E. 2020;**102**:062402

[40] Ekobena Fouda HP, Tabi CB, Mohamadou A, Kofané TC. Intramolecular vibrations and noise effects on pattern formation in a molecular helix. Journal of Physics: Condensed Matter. 2011;**23**:375104

[41] Tabi CB, Dang Koko A, Oumarou Doko R, Ekobena Fouda HP, Kofané TC. Modulated charge patterns and noise effect in a twisted DNA model with solvent interaction. Physica A. 2016;**442**:498

[42] Tabi CB, Maïna I, Mohamadou A, Ekobena Fouda HP, Kofané TC. Wave instability of intercellular Ca^{2+} oscillations. EPL. 2014;**106**:18005

[43] Tabi CB, Tagwo H, Tiofack CGL, Kofané TC. Pure quartic modulational instability in weakly nonlocal birefringent fibers. Optics Letters. 2022; **47**:5557

[44] Kraych AE, Suret P, El G, Randoux S. Nonlinear evolution of the locally induced modulational instability in fiber optics. Physical Review Letters. 2019; **122**:054101

[45] Cavalcanti SB. Modulation instability in the region of minimum group-velocity dispersion of single-mode optical fibers via an extended nonlinear Schrödinger equation. Physical Review A. 1991;**43**:6162

[46] van der Veen J. Analytic spectrum of multifrequency Raman generation with chirped pulses. Physical Review A. 2023; **108**:033516

[47] Zhang J, Donaldson W, Agrawal GP. Experimental observation of a Raman-induced temporal waveguide. Physical Review A. 2023;**107**:063518

Chapter 3

Optical Fiber Sensors

Uzma Hira and Muhammad Nayab Ahmad

Abstract

Optical fiber sensors have become an indispensable technological advancement due to their exceptional sensitivity, resilience against electromagnetic interference, and durability under challenging conditions. Their uses cover a wide range of industries, including environmental sensing, structural health monitoring, and medical diagnostics. Their performance has been improved by developments in materials, computation of signals, and miniaturization. Future developments will tackle present issues like high prices and interference from the environment by concentrating on hybrid systems, versatile capabilities, and quantum sensing. Wider acceptance will be made easier by standardization and compatibility. This chapter highlights the revolutionary power of optical fiber sensors across sectors by examining their technological developments, prospective innovations, and future possibilities.

Keywords: optical fiber sensors, structural health monitoring, environmental sensing, miniaturization, hybrid systems

1. Introduction

A flexible, translucent, cylindrical waveguide with a diameter marginally thicker than a human hair that is composed of plastic or silica is called an optical fiber. With minimal loss, optical fibers can navigate and deliver light over great distances. A translucent core with a lower refractive index is covered by a transparent cladding material to form single-index optical fibers. The whole internal reflection allows guiding light to travel through the fiber core; however, waveguide analysis shows that part of the transmitted light energy is trapped in the cladding as evanescent waves. When light hits the core–clad contact at angles larger than the critical angle, it is totally reflected and guided within the fiber [1].

Optical fiber technology, which provides fast, reliable, distant communication capabilities, is the cornerstone of current data transfer and telecommunications networks. The fundamental principle of optical fiber technology is the transmission of light signals carrying data over great distances with little loss or interference through the use of tiny glass or plastic strands. Since the 1960s, optical fiber sensors have developed in parallel with advances in optical fiber technology. The earliest sensors were straightforward and relied on changes in the fiber characteristics brought about by outside factors like strain or temperature. More advanced methods, such as fiber interferometry, were created by researchers in the 1970s, allowing for accurate measurements of physical properties [2]. As fiber optics manufacturing and signal processing advanced in the 1980s and 1990s, many types of sensors, including distributed sensors and fiber Bragg gratings, were commercially available. Optical fiber sensors gained widespread acceptance by the early 2000s in a variety of industries, including aerospace [3], medical, and telecommunications, because of their exceptional sensitivity, resistance to electromagnetic interference, and robustness under challenging conditions [4]. Future sensor applications will be increasingly more varied and sophisticated because of the focus on miniaturization, integration with new technologies like the Internet of Things, and exploration of novel sensing principles like plasmonic and photonic crystal fibers in recent years [5].

Optical fiber sensors are electromagnetically inactive. This property allows optical sensors to be used in environments with high and variable electric fields and explosive hazards, when other sensors cannot. Silica compound, the primary conduction material of optical fiber, is resistant to most chemical and biological agents and may be used in various materials and conditions. Another feature of optical fiber sensors is their minimal weight and small size [6]. Optical fiber sensors were first designed as pointwise sensors, transmitting environmental characteristics from a single site down the fiber. Numerous single-point fiber sensor topologies, including fiber Bragg grating (FBG), Inline fiber interferometers, and many more, are less expensive and easier to query. Researchers studying fiber optic sensors emphasized the potential for effective multiplexing schemes as one of the main benefits over rival technologies [1]. Fiber optic sensor multiplex reduces the cost of inquiry significantly per sensor by enabling the use of one data source and detection device for a variety of separate fiber optic sensors. Carrier frequency division, spatial-division multiplexing, optical wavelength division, time division, and coherence-domain multiplexing are a few popular multiplexing configurations that can be employed. Compared to traditional pointwise fiber sensing devices, the multiplexing schemes and signal processing techniques become more complex during the procedure of interrogation. Nevertheless, there are several benefits to developing such systems, including lower costs and higher-quality measurement outcomes for a variety of uses for sensors.

Even when multiplexed to create quasi-distributed fiber sensors, single-point fiber sensors are not able to give the same amount of information as naturally distributed optical fiber-based methods that operate throughout the whole fiber length [7]. Innovations in the optoelectronics and fiber optic communication sectors have driven down the cost of optical assemblies gradually, and the widespread accessibility of inexpensive, long-length fibers makes them useful for dispersed and remote sensing applications [8].

2. Principle

The working principle of optical fiber sensors is that they use total internal reflection to allow light to pass through the fiber core and interact with the surroundings to detect changes in parameters like pressure, temperature, strain, or composition of chemicals. These alterations modify the guided light characteristics, which are subsequently measured to determine the surrounding circumstances. Optical fiber sensors are useful instruments in many sectors for real-time surveillance and measuring applications because of their sensitive and adaptable sensing capabilities, which are made possible by the concepts of total internal reflection and light modulation. It can be simply elaborated as light can either be transmitted or reflected when it comes into contact with an interface between two media that have differing refractive indices. The outside layer of an optical fiber, called the cladding, has a lower refractive index than the core, which is where light flows. Light is fully internally reflected at the core-cladding contact. Total internal reflection keeps light within the core, enabling for effective transmission over long distances with minimal loss. When light travels as electromagnetic waves, the electric and magnetic fields fluctuate perpendicularly along the transmission path. These waves are confined and guided down the optical fibers length by the core, which serves as a waveguide. Depending on several parameters including the fiber diameter and refractive index profile, light can travel through multiple modes.

3. Types of optical fiber sensors

3.1 Fiber Bragg grating (FBG) sensors

FBGs are compact, versatile, and tiny sensing elements that provide all the advantages of optical fiber sensors. FBG sensors may be readily multiplexed in multi-point sensing networks due to the fact that the information being measured is stored in the structure's wavelength, which is an absolute feature.

3.1.1 Working

The FBG sensor uses UV light to alter the refractive index of an optical fiber core, providing spectrally controlled reflective properties. This process is highly sensitive to extension and temperature variations, making it ideal for detecting amplitude or intensity variations. This technique imparts spectrally regulated reflecting characteristics to the UV-light-treated fiber. **Figure 1** demonstrates that the reflected wavelength is highly sensitive to temperature and extension fluctuations. The sensors are integrated into the fiber's light-guiding core and are wavelength-encoded, eliminating the need for external sources [9].

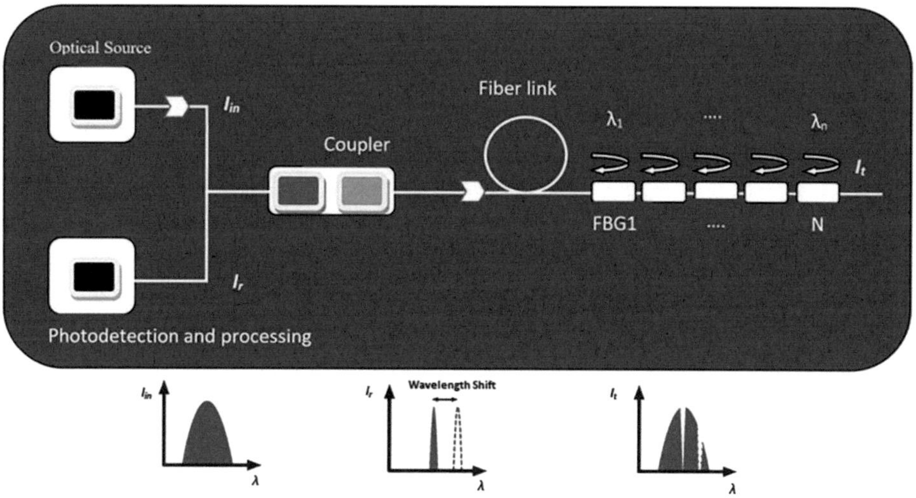

Figure 1.
Demonstration of working principle of FBG sensors.

3.1.2 Fabrication

Over the last several decades, the manufacturing process for FBGs has undergone ongoing enhancements to facilitate the effective functioning of FBGs in challenging environments characterized by very high temperatures, such as the oil and gas industries and aircraft engines, among other sectors. Currently, there has been an investigation into the use of femtosecond laser technology for the creation of microstructures, such as Fiber Bragg Gratings (FBGs). This manufacturing technique allowed FBG sensors to function at very elevated temperatures. For instance, a type II Fiber Bragg Grating (FBG) created on a standard single-mode fiber demonstrated excellent thermal stability at 1000°C for several hundred hours. Additionally, an FBG constructed using pure-silica photonic crystal fiber (PCF) proved capable of functioning at around 1300°C.

In FBG sensors, the Bragg wavelength (λ_B) or reflected light wavelength is:

$$\lambda_B = 2n_{eff}\Lambda \tag{1}$$

The grating period is denoted by Λ, whereas the effective refractive index of the fiber core is represented by n_{eff}. Changes in factors such as strain impact the grating period, whereas changes in temperature cause variations in the effective refractive index.

To evaluate individual physical parameters in FBGs, it is necessary to distinguish between the impacts of temperature and strain. Utilizing reference grating offers a pragmatic and simple method to distinguish the influences induced by temperature and strain. FBG sensors need the use of demodulators, which are sometimes referred to as interrogators, to retrieve information from the light signals sent by the sensor heads. Interrogators are responsible for interpreting variations in the Bragg wavelength, which encodes the information, and providing the measured data [10]. **Figure 1** depicts the interrogator of an FBG sensor.

3.2 Interferometric sensors

Fiber optic interferometers use the phenomenon of interference between two beams that go via distinct optical channels inside a single fiber or separate fibers. These devices need the use of beam splitting and beam combining components and are susceptible to external disturbances. By combining spectral and temporal data, interferometers can quantify changes in bandwidth, phase, intensity, frequency, and wavelength. They offer high performance, accuracy, and sensitivity. There is a current movement to make fiber optic interferometers smaller for use in micro-scale applications. This involves replacing typical bulk optic components with tiny fiber devices. Inline devices with dual optical channels inside a single physical line are very suitable for tiny interferometers owing to their inherent benefits including straightforward alignment, superior coupling efficiency, and enhanced stability [11].

3.2.1 Types of interferometric fiber optic sensors

There are three types of fiber optic interferometers: Fabry-Perot, Mach-Zehnder, and Michelson interferometer sensors.

3.2.2 Fabry-Perot interferometer sensor

A Fabry-Perot interferometer (FPI) is an optical instrument composed of two parallel reflecting surfaces that are spaced apart by a certain distance, sometimes referred to as an etalon. The interference occurs as a result of the overlapping of reflected and transmitted beams at certain surfaces. FPI sensors may be categorized as either extrinsic or intrinsic. The extrinsic FPI sensor employs reflections from an exterior cavity created by a fascinating fiber to acquire precise interference signals. This simple fabrication method requires no high-cost equipment but has disadvantages such as low coupling efficiency, careful alignment, and packaging problems. The extrinsic structure is useful for obtaining fine interference signals and requires careful alignment. Intrinsic FPI fiber sensors have reflective elements embedded into the fiber, enabling intrinsic Fabry-Perot interference. The local cavity of these sensors can be formed through various methods such as fiber Bragg gratings (FBGs), thin film deposition, micromachining, and chemical etching, forming a unique and efficient fiber sensor system [12]. **Figure 2** depicts the schematic representation of extrinsic and intrinsic FPI fiber sensors.

3.2.3 Mach-Zehnder interferometer sensors

Mach-Zehnder interferometers (MZIs) are highly adaptable sensors used in a wide range of applications due to their varied designs. Initially, Mach-Zehnder interferometers (MZIs) consisted of two separate and distinct components: the reference arm and the sensing arm. **Figure 3** shows the working reference and sensing arm. The

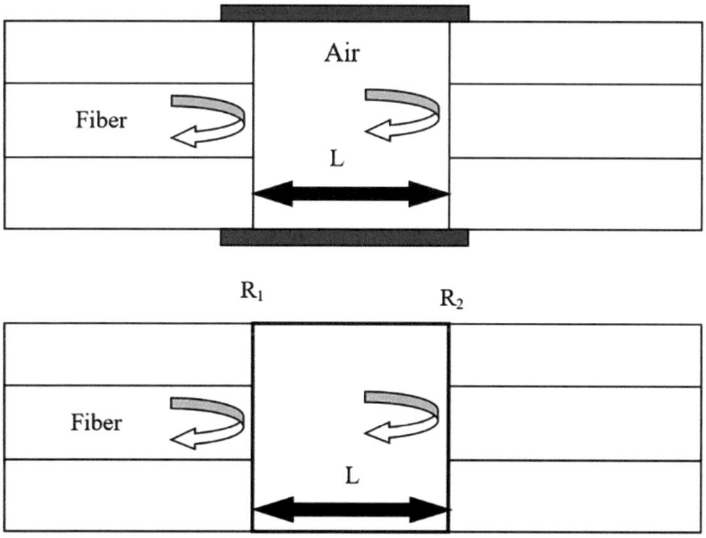

Figure 2.
(a) Extrinsic FPI sensor, and (b) intrinsic FPI sensor.

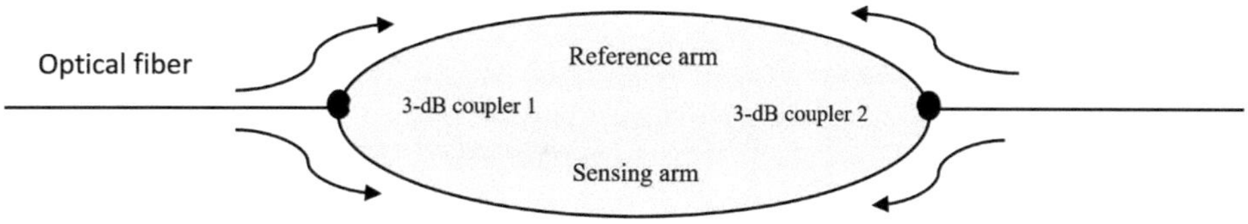

Figure 3.
An MZI (Mach-Zehnder interferometer) consists of two beam arms, namely the reference arm and the sensing arm, which are linked by means of two fiber couplers.

combined light exhibited interference due to the interaction between the two arms. In sensing applications, the reference arm is shielded from external fluctuations, whereas the sensing arm is subjected to changes. Analyzing the interference signal allows for easy detection of the changes in the sensor arm caused by temperature and strain [13].

3.2.4 Michelson interferometer sensors

Michelson interferometers (MIs) are fiber optic sensors that have similarities with MZIs in terms of their manufacturing process and operational concept. Mach-Zehnder interferometers (MIs) are optical devices that exhibit interference between beams in two separate arms. In this setup, each beam reflects at the end of its respective arm, as seen in **Figure 4**. They are compact and practically fit with multiplexing capability for parallel sensor connections [13]. It is crucial to align the difference in fiber length between the reference arm and sensing arm to match the coherence length of the light source. The mirror in the figure reflects the beams. An alternative configuration of MI, known as inline configuration, may be seen in **Figure 4b**. In this setup, a portion of the core mode beam is linked to the cladding mode(s) and then reflected by the reflector located at the end of the fiber [14].

3.3 Chemical and biosensors

The primary use of fiber optic sensors is in the field of biology and medicine, namely within the visible and near infrared spectral range, which spans from 340 nm to 2 μm. Fibers composed mostly of silicon dioxide (SiO2) are appropriate for this specific range of wavelengths. The available types of fibers include single mode (SM) and multimode (MM), which may be further classified as Step Index (SI) fibers like PCS (polymer cladding silica) fibers, or Gradient Index (GI) fibers, or fiber bundles made from these fibers [15]. **Figure 5** has shown system for detecting environmental estrogens in which surface plasmon resonance-based TFBG sensing mechanism has been used.

Hollow fibers, sectorial fibers, U-shaped fibers, inverted gradient index profiles, down or up tapered fibers, and PCF fibers (polymer-clad microstructure fiber) are some of the specially designed fibers that are presently commercially manufactured. The purpose of developing and studying these optical elements (OFs) is to enhance the evanescent field, which leads to improved sensitivity and detection limits of FOS. Two-cone tapered fibers were used to determine chemical concentrations in transmission arrangements and develop tapered fiber optical elements. A latest variant of a fiber optical element (FOE) known as V-taper has been created specifically for fiber

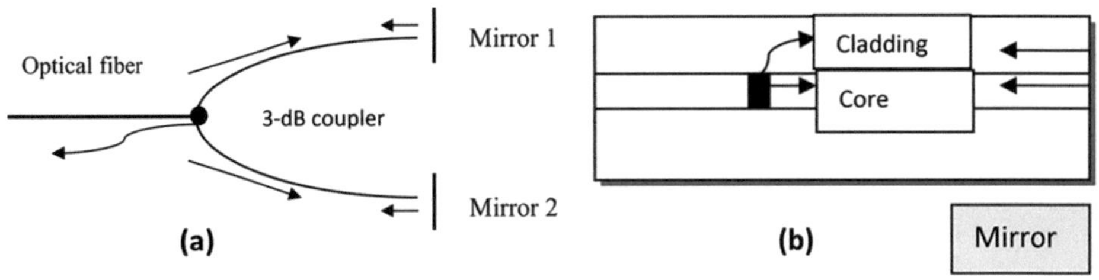

Figure 4.
(a) Schematic representation of Michelson interferometer, and (b) inline Michelson interferometer.

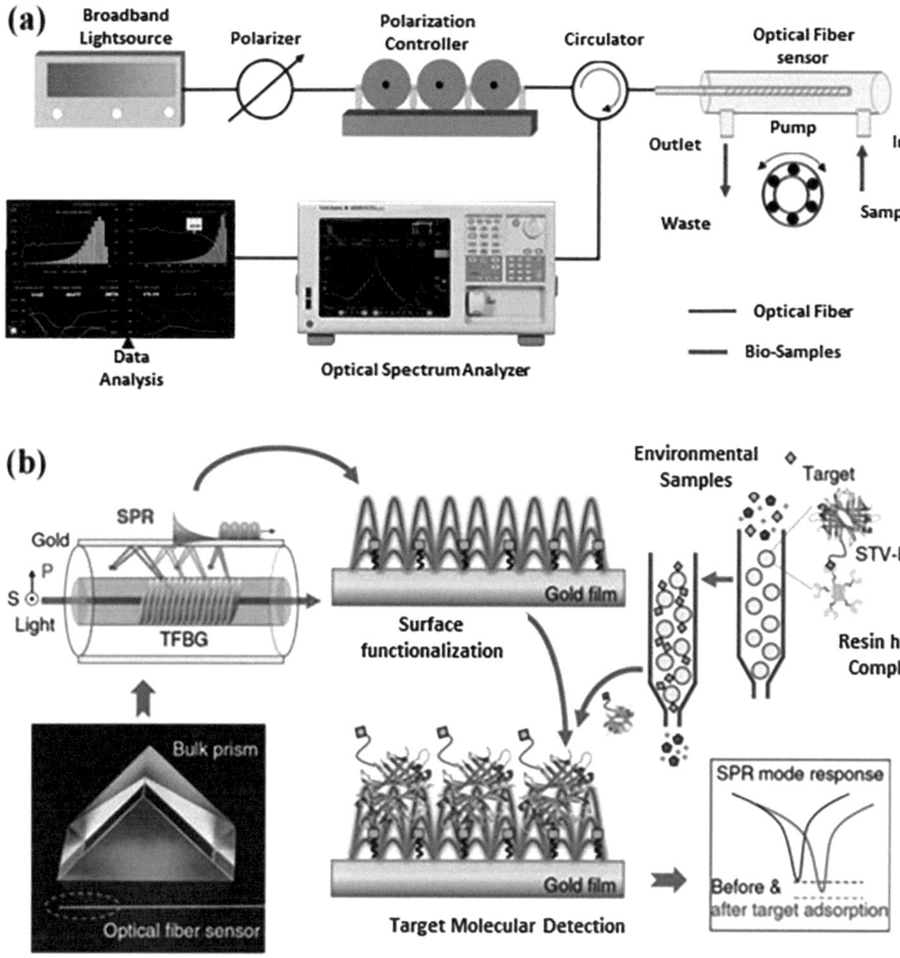

Figure 5.
(a) A system for detecting environmental estrogens, (b) Surface plasmon resonance-based TFBG sensing mechanism for the identification of ambient estrogens in water.

optic chemical sensors (FOCS) in reflection configurations. This FOE is designed for precise measurements in human, animal, and plant cells, as well as tiny quantities of specialized liquids [16].

3.3.1 Fiber optic chemical sensor (FOCS)

The FOCS is capable of detecting various gases such as hydrogen, hydrocarbons, oxygen, hydrogen peroxide, ozone gas, sulfur and nitrogen oxides, ammonia, chloride ions, H^{+}, Ca^{2+}, Ag^{+}, Fe^{3+}, Zn^{2+}, as well as specific chemical compounds used for environmental measurements, including pollutants, agrochemical agents, drugs, and pharmaceuticals [17]. The FOCS systems typically consist of three fundamental components:

(1) An active sensing part, such as a special optic fiber, uses a transducer with immobilized reagents to recognize molecular/ionic elements, producing an optical signal that is then sent to the detection system. (2) A detection system uses optical signal parameters like intensity, frequency, and phase to convert monitoring parameters like refractive

index, pH, and chemical concentrations into useful electrical signals like current, voltage, or frequency. (3) A control and evaluation system is a computer control and software that allows users to measure data in real time and online [18]. **Figure 6** shows the components and application domains of fiber optic chemical sensors and biosensors.

3.3.2 Fiber optic biosensors (FOBS)

Based on their analyte interaction, FOBS may be affinity or catalytic, and their biological recognition components can be classified into four categories.

1. Enzyme optical biosensors use enzymes to facilitate reactions with high specificity. The resulting products or the depletion of substrates are detected either directly or by coming into contact with an indicator.

2. Whole-cell optical biosensors examine the impact of an analyte on a whole cell, often a microbial cell. Optical detection refers to the identification of either a specific condition (such as pH or oxygen levels) or a characteristic of the cells themselves, such as the generation or disappearance of fluorescence or bioluminescence.

3. Immunoassay optical biosensors use the particular interaction between an antibody and an antigen for detection. Binding may be detected by the use of a fluorescent label or by monitoring a change in refractive index or reflectivity, which does not need the use of a label.

4. Nucleic acid optical biosensors use the binding properties of single-stranded DNA (ssDNA) to produce double-stranded DNA. Usually, these sensors need the labeling of one strand of single-stranded DNA (ssDNA) with an optical marker [19, 20]. **Figure 7** illustrates the incorporation of OFB in microfluidic chips and shows the construction and working of biosensors.

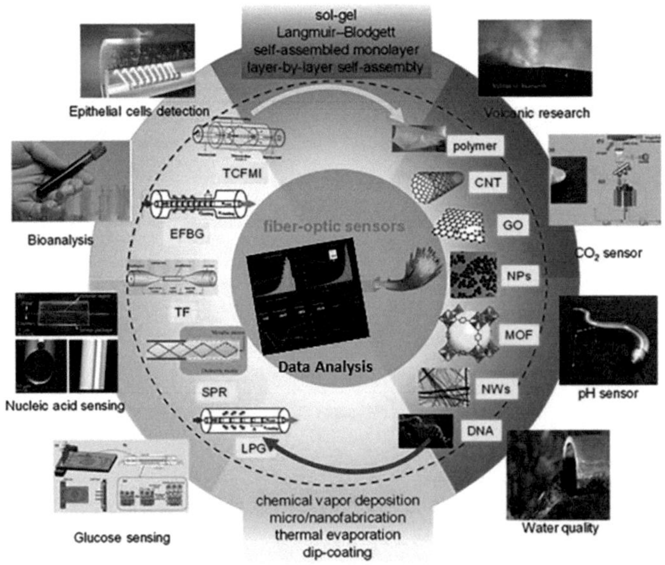

Figure 6.
Components and diverse applications of fiber optic chemical sensors and biosensors.

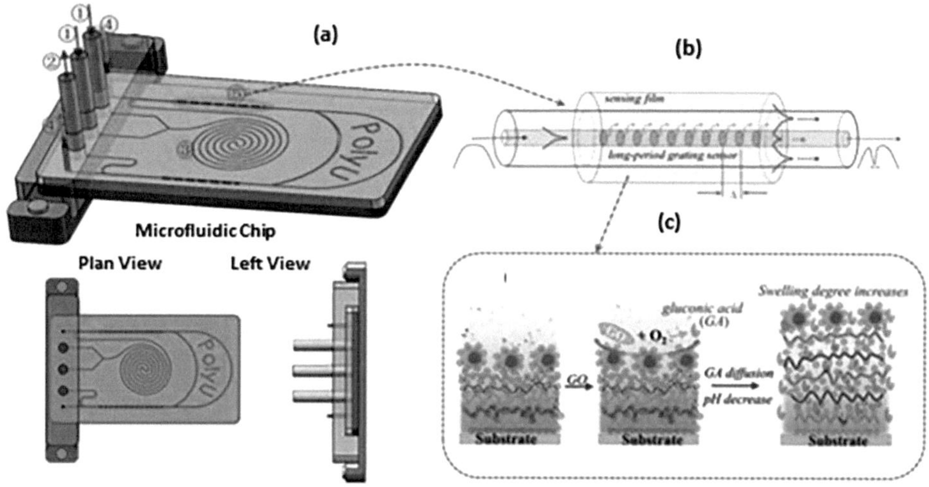

Figure 7.
(a) Schematic representation of a microfluidic chip that incorporates an optical fiber biosensor, (b) illustration of the construction of the biosensor, and (c) demonstration of the method by which the biosensor senses biofilm.

3.4 Intensity-based sensors

These sensors detect changes in light intensity based on the surrounding environment. Transmission, reflection, and microbending are three phenomena linked with intensity modulation. A reflecting or transmissive target can be included within the fiber to achieve this. Absorption, scattering, fluorescence, and polarization are other processes that may be used individually or in conjunction with the three primary concepts. Intensity-modulated sensors often need a greater amount of light compared to phase-modulated sensors, which in turn necessitates the usage of bigger multimode fibers or bundles. Enzyme optical biosensors use enzymes to catalyze reactions that have excellent specificity; products or substrate consumption are detected directly or by interacting with an indicator [21].

3.5 Distributed temperature sensors (DTS)

Optical fibers have been used as temperature sensors for several years, using absorption, scattering, fluorescence, index variations, and thermal emission. These sensors have a passive function of light guiding, while others use temperature-dependent parameters like phase delay and polarization to modify light propagation. These sensors measure temperature in a single point, allowing for sensitive and sophisticated interferometric arrangements to detect very little temperature shifts. The sensor detects temperature-induced backscatter power variations at numerous fiber sites using optical time-domain reflectometry (OTDR) [22].

The sensor relies on the temperature sensitivity of the refractive index differential between the silica core and organic cladding in plastic-cladding-silica (PCS) fibers, which may cause disruptions in transmission when there are variations in temperature. To improve efficiency, bend losses are introduced under a strong curvature, and radiation losses decrease with the index difference and temperature. As the value of n_1-n_2 increases from −20° to 60°, good sensitivity can be predicted over a wide range.

Fiber-distributed temperature sensors are used to monitor electrical power cables in order to detect damaging overheating and identify hot spots. For this reason, fiber was bundled in the screen layer of a 60 kV cable, with the fiber's excellent dielectric characteristics enabling light propagation insensitive to the wire's severe electromagnetic environment.

Another application is the detection of concrete setting temperature in civil engineering works [23].

3.6 Distributed strain sensors (DSS)

Measuring the distribution of strain is essential for evaluating the state and monitoring the structural health of materials and engineering constructions [24]. The system has the ability to detect, categorize, pinpoint, and measure abnormalities such as fractures in concrete and steel, deterioration of steel bars embedded in concrete, and self-induced contraction of ultra-high-performance concrete. Technology such as fiber optic sensors, vibrating wire strain gauges, and digital image correlation emerged from the need to quantify strain in the process of developing and characterizing materials.

Fiber optic strain sensors have several benefits compared to conventional strain sensors, such as resistance to electromagnetic interference, small dimensions, low weight, exceptional precision, heightened sensitivity, and excellent chemical and physical durability.

Optical fibers with protective coatings have been used as distributed strain sensors for automated inspection of various engineering structures during their construction, operation, and maintenance. A point sensor is limited to providing a single measurement at a specific location along its gauge length. On the other hand, a distributed sensor is capable of providing measurements that are spread out throughout the whole length of the optical fiber. This makes distributed sensors suitable for automated health surveillance of substantial engineering systems [25]. Fiber optic sensors are equipped with protective coatings to enhance their mechanical durability when inserted into or connected to a host matrix for the purpose of monitoring strain. The stresses experienced by fiber optic sensors may vary from the strains present in the host matrix as a result of the presence of protective coatings. This phenomenon is referred to as the strain transfer effect. Fiber optic sensors are used to quantify stresses at a particular location, and a strain transfer ratio is established to characterize the extent of strain transfer. This ratio is equivalent to the strain measured divided by the actual strain in the host matrix. The ratio is contingent upon the packing technique, the substance used for coating, the diameter of the coating, and the length of bonding. Point sensor strain transfer ratios often fall within the range of 0.90–0.96. It is possible to calibrate a constant ratio for each sensor to compensate for the sensor's strain transfer impact [26]. The strain transfer effect is shown in **Figure 8**.

3.7 Current and magnetic field sensors

Optical fiber sensors that detect magnetic fields and electric currents, using the Faraday effect, have been used for many years. Increasing the number of fiber coils may boost the sensitivity of these sensors, but this leads to increase linear birefringence, limiting sensitivity and affecting response linearity. Spun optical fibers can achieve a large current measurement range and 0.1% accuracy. Both polarimetric

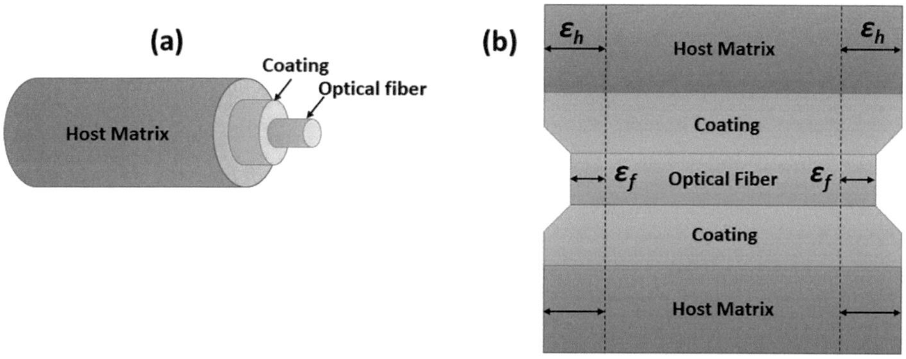

Figure 8.
Strain transfer effect (a) optical fiber packed in host matrix, and (b) strains in host matrix and optical fiber.

and interferometric detection arrangements provide benefits, but they also come with drawbacks such as intricate optical setups, expensive components, and complex questioning methods. To address issues with magnetic field measurements, various approaches have been tested, including using magnetostrictive, magnetoptical, and magnetic fluid materials in combination with optical fibers [27]. Magnetostrictive transducers use Fiber Bragg gratings (FBG) to identify distortions caused by magnetic fields, therefore transforming observations into spectral regions. Magnetoptical materials such as Yb Bi: YIG, BGO crystals, TG20 paramagnetic glass, and SF2 are used in conventional polarimetric setups that include laser or LED sources, polarizers, analyzers, and photodetectors. Magnetostrictive fluids are integrated with a Fabry-Perot interferometer and a broad-band source to transform observations into spectral domains. To accurately measure magnetic fields ranging from 50 to 60 mT with an error margin of less than 2%, specialized photonic crystal fibers and optical fiber Fabry-Perot interferometers, tapers, or gratings are necessary. These sophisticated sensors, which operate in the spectral domain, are still a distant goal of commercialization. $Bi_{12}SiO_{20}$ (BSO) crystals, including undoped and Cr, Mn, Cu doped BSO, and $Bi_{12}SiO_{20}$: Ni, are magnetoptical materials with high Verdet constants [28].

3.8 Displacement and position sensors

Fiber optic sensors have become more popular for measuring displacement because of their compact probe sizes and non-contact characteristics, which make them well-suited for use in hazardous environments. Techniques used for these measurements including optical interferometry, wavelength modulation, and reflective intensity modulations. Laser interferometry, which relies on fringe counting, provides excellent precision and consistency. However, it is reliant on the stability of light wavelength, which may lead to higher expenses. Wavelength modulation necessitates the use of an optical spectrum analyzer to identify wavelengths associated with changes in power or current, which may be expensive. Intensity modulation is a straightforward method used in displacement sensors to compare the intensity of reflected light with the intensity of the light that is first emitted. Plastic fibers with large core radii and high numerical aperture, known as multimode fibers, are extensively used because of their exceptional ability to couple signals and gather the greatest amount of reflected light from the target [29, 30].

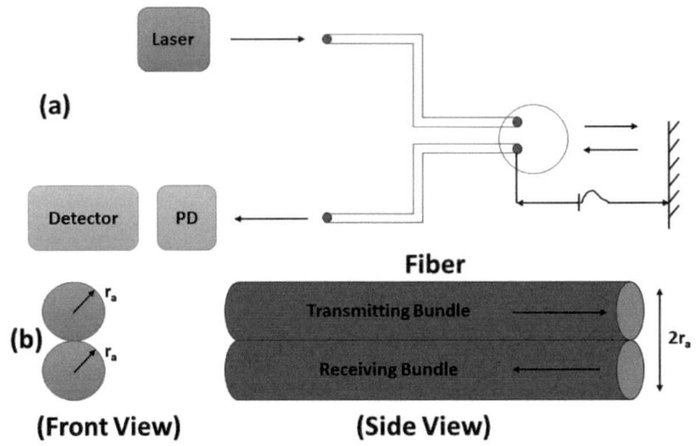

Figure 9.
Basic FODS system setup (a) system configuration, and (b) sensor probe front and side views.

3.8.1 Sensing principle

In a conventional FODS (fiber optic displacement sensor) system, a sensor probe is often used beside a flat reflecting target. The transmission and receiving fibers of the pair are consolidated. The figure illustrates the fundamental configuration of the FODS system. The sensor shown in **Figure 9(a)** comprises a photodiode detector, a fiber optic probe, and a light source. The sensor probe in **Figure 9(b)** consists of two parallelly bundled plastic optical fibers (POFs) that are similar in nature, as seen from both the front and side perspectives.

The functionality of a fiber optic distributed sensing (FODS) system is greatly impacted by many components, such as the intensity of the ambient light, the placement of the optical fiber bundle and reflecting surface, and the reflectivity of the surface. When developing FODSs, achieving high sensitivity and a wide linear range is often done via two categories: designing the sensor probe and designing the target reflector. When choosing a laser source, it is important to take into account factors such as cost, operating wavelength, output power, and stability [31].

4. Applications

Optical fiber sensors are available in a variety of types, each tailored to a specific sensing application. To demonstrate the adaptability of optical fiber sensors, several applications are considered, including environmental tracking, biological sensing, industrial process management, security and protection, and infrastructure surveillance. **Figure 10** shows OF with different applications.

4.1 Biomedical sensors

Biomedical optical fiber sensors are small devices that doctors can use to test anything within your body, such as your blood sugar or how acidic something is, without having to stab you with needles. They function by passing light via optical fibers and detecting how it changes, and they may one day be used in wearable gadgets to monitor your health constantly [32]. Biomedical optical fiber sensors have the potential to be more compact and ergonomic in the future, resulting in improved use and

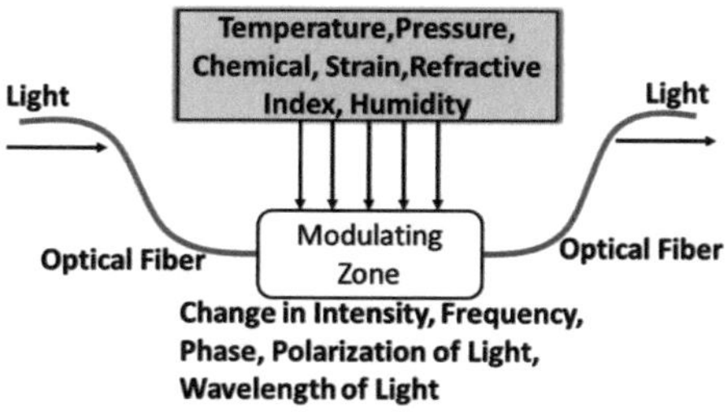

Figure 10.
Optical fiber sensor with different applications.

comfort. They might also be able to measure even more things, helping doctors and individuals monitor health in real time, leading to more personalized and proactive healthcare approaches. Additionally, advancements in data analysis and connectivity could enable seamless integration with digital health platforms for comprehensive health management [33].

Pulse oximeters for measuring blood oxygen levels, glucose sensors for monitoring blood sugar, and pH sensors for determining acidity levels in physiological fluids are the most popular biomedical optical sensors used in hospitals [34]. In addition, optical coherence tomography (OCT) enables the visualization of tissues with a high level of detail, while fluorescence-based sensors can detect biomarkers for a variety of disorders. These sensors serve important roles in diagnosis, therapy monitoring, and patient care in a variety of medical specializations.

4.2 Environmental monitoring sensors

Optical fiber sensors are vital tools for environmental monitoring in a variety of sectors. In the atmosphere, gas sensors constantly track gases such as CO_2, CH_4, and VOCs, assisting in thorough air quality evaluation and pollution reduction efforts. They act as attentive monitors of water quality, quickly identifying toxins such as toxic chemicals, heavy metals, and organic contaminants in rivers and lakes, allowing for swift actions to protect ecosystems. pH sensors give critical information about water acidity or alkalinity levels, allowing for a more detailed analysis of aquatic wellness as well as regulatory compliance. Particulate matter sensors supplement these efforts by thoroughly quantifying PM-10 and PM-2.5 concentrations, providing insight into air pollution concentrations, and directing efficient reduction tactics. Furthermore, temperature and humidity sensors meticulously track environmental factors such as ambient conditions and soil moisture content, providing significant insights into climate change, agriculture, and soil conservation methods. Oceanographic sensors, located beneath the ocean's surface, monitor seawater salinity, ocean depths, and earthquakes [35], providing critical data for marine research, emergency preparation, and marine conservation [36]. These optical fiber sensors constitute a strong network, working hard to conserve and defend our beautiful environment for future generations.

4.3 Industrial sensors

Optical fiber sensors are important in industrial applications because of their great sensitivity, endurance, and flexibility. These sensors work by delivering light over optical fibers to detect variations in ambient factors including pressure, temperature, and vibration. This capacity is particularly valuable in areas such as production, where accurate measurements are required for quality assurance and efficiency. Optical fibers are non-conductive and, hence, appropriate for use in explosive or high-voltage settings, assuring safety and dependability. Furthermore, their durability against rust and severe temperatures increases their life expectancy and decreases maintenance costs [2].

Within the energy industry, they keep an eye on the structural integrity of oil pipelines and wind turbine blades, looking for signs of stress and possible problems before they happen. In the automobile sector, these sensors are utilized to monitor tire pressure and engine performance in real time, assuring both safety and economy. In civil engineering, these sensors are implanted in bridges and buildings to monitor load and identify early structural deterioration. They are used in the pharmaceutical sector to precisely track temperature and chemical levels during medication production procedures. In the food and beverage business, temperature and pressure are monitored in processing and storage areas to guarantee quality control. Optical fiber sensors are also used in mining to monitor ground changes and vibrations, which improve worker safety [37].

5. Development in optical Fiber sensor technology

5.1 Enhanced signal processing techniques

The reliability of optical fiber sensors has been boosted by developments in signal processing strategies, such as digital screening, noise elimination algorithms, and sophisticated demodulation techniques. These improvements make it possible to read information more reliably and accurately, which is crucial for applications in industrial automation, healthcare diagnosis, and telecommunications.

5.2 Improved precision and sensitivity

Enhancing the sensitivity and precision of optical fiber sensors has been the primary objective of recent advancements. The employment of high-reflectivity gratings and sophisticated writing processes, among other innovations in fiber Bragg grating (FBG) technology, have made it possible to detect strain and temperature with greater accuracy. Additionally, exceptional precision for a variety of applications has been made possible by the fabrication of ultra-sensitive interferometric sensors utilizing cutting-edge interferometry methods.

These sensors are suited for industries like aerospace and biomedical engineering that need ultra-accurate evaluations because they further improve their accuracy using advanced algorithms for signal processing like phase demodulation and coherence sensing [38].

5.3 Integration and miniaturization

Micro- and nanostructured optical fibers have been developed as a result of the tendency toward miniaturization. These little sensors have applications in wearable gadgets, medical instruments, and surveillance of structural health. They may be placed in different materials and combined into small devices. Technological developments in manufacturing, including femtosecond laser writing, have made it possible to produce extremely complex fiber structures with enhanced performance properties.

Micro-electromechanical systems (MEMS) and optical fibers have been combined to create integrated, small-sized detectors. These hybrid systems provide new opportunities in areas such as environmental surveillance and healthcare diagnosis by combining the advantages of MEMS's compact size and elasticity with the high sensitivity of optical fiber sensors.

5.4 Remote and wireless sensing

Wireless technological advancements have made it possible to create distant optical fiber sensor devices. These devices transfer sensor data from distant locations to central stations for monitoring via wireless communication techniques. This feature is crucial for applications like environmental monitoring, emergency management, and military inspection, when standard cable connections may not be feasible [39].

5.5 Enhanced materials and coatings

The performance of optical fiber sensors has increased with the introduction of specialty optical fibers, such as those made of unique materials like chalcogenide glass or loaded with rare-earth elements. These fibers have increased endurance under challenging conditions, increased sensitivity, and longer operating wavelength ranges. Teflon, gold, and polyimide are a few examples of coatings that offer defense against harsh weather conditions, chemicals, pollutants, and moisture. These coatings extend the lifespan and durability of optical fiber sensors, enabling their application in demanding environments such as those seen in the chemical, aerospace, and oil and gas sectors.

5.6 Smart sensing and data analytics

Optical fiber sensors that are connected to IoT platforms allow for real-time data collecting and analysis through communication with other devices and systems. This is known as smart sensing and data analytics. Proactive upkeep, improved decision-making, and more effective monitoring and management of industrial operations are all made possible by this connection. Advanced patterns, anomaly identification, and forecasting may be achieved by applying artificial intelligence as well as machine learning to the data produced by optical fiber sensors. These technologies increase operational efficiency and decrease downtime by improving the capacity to analyze complicated data sets [40].

5.7 High-temperature and harsh environment sensors

The range of applications for optical fiber sensors has been extended by specialized optical fibers made to endure extreme radiation levels, corrosive conditions, and severe temperatures. These sturdy sensors are crucial for keeping an eye on conditions in sectors like deep-sea research, nuclear energy, and space exploration.

5.8 Eco-friendly and sustainable solutions

Sustainable development is becoming more and more important in optical fiber sensor technologies. Utilizing recyclable parts, energy-efficient designs, and environmentally friendly materials helps to save the environment and advances global sustainability objectives. Optical fiber sensors support green technology initiative in many different sectors of the economy.

5.9 Cost reduction and commercialization

Advancement in manufacturing processes and economies of scale has reduced the cost of optical fiber sensors, making them more economically feasible. The technology is now more affordable and suitable for a larger range of uses, including consumer electronics and industrial automation. The introduction of standardized protocols and interfaces has also aided in the commercialization of optical fiber sensors by simplifying their integration into current systems [19].

6. Challenges and future trends

6.1 Challenges

Sensitivity and precision are influenced by external interference such as changes in water content, mechanical stress, and temperature. Signal attenuation and loss can happen over considerable distances, which lowers the sensors efficacy and accuracy. Ensuring signal integrity is essential for precise data transfer. In order to analyze complex sensor data in real time, advanced signal processing methods are needed. Cross-sensitivity problems might arise from optical fiber sensors' simultaneous response to several factors. It is difficult to develop sensors that can identify particular characteristics selectively and interference-free. Precise manufacturing quality control is essential since deviations might impair output. It is difficult to maintain great sensitivity in smaller forms. Developing, installing, and maintaining optical fiber sensors may be expensive, which prevents their broad use, particularly in settings with limited resources.

The enormous volume of data produced by optical fiber sensors makes complex processing and interpretation methods necessary, which makes real-time evaluation and selection difficult. It is crucial to maintain constant efficiency and high standards throughout the production process. Variations can affect the sensitivity and dependability of sensors, which is why precision production is crucial. The integration of optical fiber sensors with current systems and infrastructure might present difficulties because of challenges with scalability and standardization. Achieving smooth integration is essential to successful deployment. It is difficult to get great sensitivity with miniaturized optical fiber sensors. Advanced engineering and production processes are needed to balance performance with small size.

6.2 Future trends

Recent trends in optical fiber design and production have led to the development of custom biosensors. These include microstructured fibers with optofluidic channels, air holes, biolayers, multicore optical fibers, and polymer optical fibers. These fibers allow for direct injection of fluids or gases, improve light propagation, and explore polarization and cross-talk effects. Despite their sensitivity and convenience, these fibers are increasingly being studied for their potential in biosensors [41].

Optical fiber biosensors are advancing toward aptamer-based technologies for diagnostics and online environmental monitoring, improving biosensing performance, particularly on optical devices with depth-sensitive fields. DNA bioreceptors offer high stability and resistance to variations in sensing conditions, making them ideal for long-term measurements and real-time, reversible sensing. Optical fiber-based technologies offer assets for regeneration and new binding possibilities, but point-of-care devices are primarily single-use. Other purposes, such as monitoring enzyme activity or water quality, might be more suited for regeneration and rely on this aptamer-based technology to extend their longevity as online biosensors. Research on this fiber has discussed various methods for detecting heavy metal ions (Pb2+, Cu2+ Hg2+, etc.) introduced into the environment [42].

Sensor sensitivity and durability can be improved by adopting cutting-edge materials such as photonic crystal fibers, nanomaterials, and complex composites. Artificial intelligence is being used as a breakthrough signal-processing approach to improve data quality and understanding in real time.Including optical fiber sensors in hybrid systems for all-encompassing monitoring, such as MEMS and electrical sensors. Using flexible electronics and miniaturization technologies to build small, light, and incredibly sensitive sensors. Encouraging homogeneity and compatibility for easier integration with existing infrastructure and wider adoption.

Developing sensors with multifunctional monitoring capabilities by developing them to detect many parameters at once. Investigating biocompatible sensor technologies for medical applications such as on-the-spot physiological observation and testing.

Creating innovative sensors for the energy sector, including renewable energy systems, pipeline tracking, and oil and gas development. Developing sensors for structure health monitoring, vibration measurement, and other applications in the aviation and automotive industries. Sensors for agricultural and food safety applications, such as contamination detection and soil quality monitoring, are under investigation. Creating innovative sensors for wearable devices like smart watches. Developing cutting-edge sensors for environmental applications such as pollution monitoring and climate change impact research. Improving sensors for microscopy and endoscopic applications that require high-resolution pictures. Developing sturdy sensors for harsh situations such as space and deep-sea research are also under study for future development.

7. Conclusion

Optical fiber sensors are a revolutionary technology with a wide range of applications in many different sectors. Their great sensitivity, resilience to electromagnetic interference, and robustness in hostile conditions make them perfect for a variety of applications, ranging from surveillance of structural health to diagnostic purposes. Recent developments in signal processing, miniaturization, and materials have greatly improved their efficiency and adaptability. Subsequent advancements are expected to further enhance their potential by integrating hybrid systems and quantum sensing for multipurpose uses. Optic fiber sensors have a bright future ahead of them despite obstacles like high prices and interference from the environment. These problems are being addressed by further research and development. By ensuring wider usage through standardization and interoperability initiatives, these sensors will become a crucial component of contemporary technological breakthroughs. Looking ahead, optical fiber sensors continue to hold enormous promise for revolutionizing a variety of industries due to their constant invention and incorporation into a wide range of applications.

Author details

Uzma Hira* and Muhammad Nayab Ahmad
School of Physical Sciences (SPS), University of the Punjab, Lahore, Pakistan

*Address all correspondence to: uzma.sps@pu.edu.pk

References

[1] Udd E, Spillman WB Jr. Fiber Optic Sensors: An Introduction for Engineers and Scientists. Hoboken, NJ: John Wiley and Sons; 2024

[2] Floris I et al. Fiber optic shape sensors: A comprehensive review. Optics and Lasers in Engineering. 2021;**139**:106508

[3] McKenzie I et al. Fiber optic sensing in spacecraft engineering: An historical perspective from the European space agency. Frontiers in Physics. 2021;**9**:719441

[4] Glisic B. Concise historic overview of strain sensors used in the monitoring of civil structures: The first one hundred years. Sensors. 2022;**22**(6):2397

[5] Liu Y, Peng W. Fiber-optic surface plasmon resonance sensors and biochemical applications: A review. Journal of Lightwave Technology. 2020;**39**(12):3781-3791

[6] Mizuno Y et al. Distributed polymer optical fiber sensors: A review and outlook. Photonics Research. 2021;**9**(9):1719-1733

[7] Ip E et al. Distributed fiber sensor network using telecom cables as sensing media: Technology advancements and applications. Journal of Optical Communications and Networking. 2022;**14**(1):A61-A68

[8] Zhang J et al. Recent advances in optical fiber enabled radiation sensors. Sensors. 2022;**22**(3):1126

[9] Zhang J et al. An optical fiber sensor based on polyimide coated fiber Bragg grating for measurement of relative humidity. Optical Fiber Technology. 2021;**61**:102406

[10] Sahota JK, Gupta N, Dhawan D. Fiber Bragg grating sensors for monitoring of physical parameters: A comprehensive review. Optical Engineering. 2020;**59**(6):060901-060901

[11] Miliou A. In-fiber interferometric-based sensors: Overview and recent advances. Photonics (MDPI). 7 Jul 2021;**8**(7):265

[12] Liu K et al. Interferometer-based distributed optical fiber sensors in long-distance vibration detection: A review. IEEE Sensors Journal. 2022;**22**(22):21428-21444

[13] Dandridge A. Fiber optic sensors based on the Mach–Zehnder and Michelson interferometers. In: Fiber Optic Sensors: An Introduction for Engineers and Scientists. Hoboken, NJ: Wiley; 2024. pp. 213-248

[14] Guo J et al. High-temperature measurement of a fiber probe sensor based on the Michelson interferometer. Sensors. 2021;**22**(1):289

[15] Cennamo N, Pesavento M, Zeni L. A review on simple and highly sensitive plastic optical fiber probes for bio-chemical sensing. Sensors and Actuators B: Chemical. 2021;**331**:129393

[16] Li L et al. Optical fiber optofluidic bio-chemical sensors: A review. Laser and Photonics Reviews. 2021;**15**(7):2000526

[17] Esposito F. Chemical sensors based on long period fiber gratings: A review. Results in Optics. 2021;**5**:100196

[18] Zhao Y et al. Applications of fiber-optic biochemical sensor in microfluidic chips: A review. Biosensors and Bioelectronics. 2020;**166**:112447

[19] Leitão C et al. Cost-effective fiber optic solutions for biosensing. Biosensors. 2022;**12**(8):575

[20] Esfahani Monfared Y. Overview of recent advances in the design of plasmonic fiber-optic biosensors. Biosensors. 2020;**10**(7):77

[21] Mitchell GL. Intensity-based and Fabry–Perot interferometer sensors. In: Fiber Optic Sensors: An Introduction for Engineers and Scientists. Hoboken, NJ: Wiley; 2024. pp. 125-137

[22] Roriz P et al. Optical fiber temperature sensors and their biomedical applications. Sensors. 2020;**20**(7):2113

[23] Kersey AD. Distributed and multiplexed fiber optic sensors. In: Fiber Optic Sensors: An Introduction for Engineers and Scientists. Hoboken, NJ: Wiley; 2024. pp. 263-291

[24] Soman R, Wee J, Peters K. Optical fiber sensors for ultrasonic structural health monitoring: A review. Sensors. 2021;**21**(21):7345

[25] Bai H et al. Stretchable distributed fiber-optic sensors. Science. 2020;**370**(6518):848-852

[26] Tan X et al. Strain transfer effect in distributed fiber optic sensors under an arbitrary field. Automation in Construction. 2021;**124**:103597

[27] Liu C et al. Applications of magneto-strictive, magneto-optical, magnetic fluid materials in optical fiber current sensors and optical fiber magnetic field sensors: A review. Optical Fiber Technology. 2021;**65**:102634

[28] Bucholtz F. Fiber optic magnetic sensors. In: Fiber Optic Sensors: An Introduction for Engineers and Scientists. Hoboken, NJ: Wiley; 2024. pp. 545-572

[29] Bonopera M. Fiber-bragg-grating-based displacement sensors: Review of recent advances. Materials. 2022;**15**(16):5561

[30] Acharya A, Kawade N. A Fabry–Perot interferometer-based fiber optic dynamic displacement sensor with an analog in-phase/quadrature generator. IEEE Sensors Journal. 2020;**20**(24):14764-14771

[31] Ashry I et al. A review of distributed fiber–optic sensing in the oil and gas industry. Journal of Lightwave Technology. 2022;**40**(5):1407-1431

[32] Bado MF, Casas JR, Gómez J. Post-processing algorithms for distributed optical fiber sensing in structural health monitoring applications. Structural Health Monitoring. 2021;**20**(2):661-680

[33] Loayssa A. Optical fiber sensors for structural health monitoring. In: New Developments in Sensing Technology for Structural Health Monitoring. Dordrecht: Springer; 2011. pp. 335-358

[34] Gong P et al. Optical fiber sensors for glucose concentration measurement: A review. Optics and Laser Technology. 2021;**139**:106981

[35] Lindsey NJ, Martin ER. Fiber-optic seismology. Annual Review of Earth and Planetary Sciences. 2021;**49**:309-336

[36] Liang H et al. Review of optical fiber sensors for temperature, salinity, and pressure sensing and measurement in seawater. Sensors. 2022;**22**(14):5363

[37] Berthold JW III. Industrial applications of fiber optic sensors. In: Fiber Optic Sensors: An Introduction for Engineers and Scientists. Hoboken, NJ: Wiley; 2024. pp. 573-591

[38] Li J. A review: Development of novel fiber-optic platforms for bulk and surface refractive index sensing applications. Sensors and Actuators Reports. 2020;**2**(1):100018

[39] Venketeswaran A et al. Recent advances in machine learning for fiber optic sensor applications. Advanced Intelligent Systems. 2022;**4**(1):2100067

[40] Du C et al. A review of railway infrastructure monitoring using fiber optic sensors. Sensors and Actuators A: Physical. 2020;**303**:111728

[41] Loyez M et al. Overview and emerging trends in optical fiber aptasensing. Biosensors and Bioelectronics. 2022;**196**:113694

[42] Schuster K et al. Material and technology trends in fiber optics. Advanced Optical Technologies. 2014;**3**(4):447-468

Chapter 4

Multimode Interference Sensors for Static and Dynamic Monitoring

Ahmed Hisham E. Morshed

Abstract

Because of their simplicity and lower cost of realization, intensity-based optical fiber sensors are of great industrial and research interest. A class of these sensors uses the interference of waves propagating in a fiber structure to induce intensity variations that are then recovered using direct detection. These in-fiber interferometers make use of the sensitive phase variations of waves propagating in fibers to produce intensity variations, resulting in better sensitivities compared to many pure intensity-based sensors. This chapter addresses simple optical fiber sensors based on modal interference in multimode optical fibers: their working principles, potential applications, and challenges for industrial sensor realizations. Different sensor structures and approaches to sensing have been developed. These are reviewed and contrasted to one another. Sensor applications and multiplexing schemes are then addressed and compared.

Keywords: optical fiber sensors, multimode interference, dynamic monitoring, strain measurement, environmental monitoring

1. Introduction

In addition to their use in transmission for communication, optical fibers have extensively been used to realize optical sensors. These sensors are very promising as optical fibers are hazard-free and immune to electromagnetic interference, and their use enables the sensors to be operated remotely, with the sensor electronics situated away from the sensing sites. They are also small in size, lightweight and can be networked by their connecting fibers, achieving sensing over a vast monitoring area using reduced wiring and cost-effective installations [1].

Optical sensors can be classified according to the way the measured quantity (also called the measurand) affects the light signal in the sensor, which dictates the type of scheme usable for sensor signal detection. If the measurand affects the intensity of light, then direct detection can be used to recover the modulating changes. In this case, we have an intensity-modulated sensor. On the other hand, if the measurand affects the phase of the light wave, an interferometric detection scheme needs to be used to recover the induced phase changes. We then have phase-modulated (also called interferometric) sensors. As the phase of a propagating wave is more sensitive to changes in the fiber conditions compared to the wave intensity, phase-modulated sensors are generally more sensitive than intensity modulated ones, but they are relatively complicated and more expensive [2].

Because of their simplicity and lower cost of realization, intensity sensors are of great industrial and research interest, and have been the subject of extensive investigations. Many fiber structures and sensor innovations have been proposed and investigated. A class of these sensors uses the interference of waves propagating in the same fiber to induce intensity variations that are then recovered using direct detection. These in-fiber interferometers make use of the sensitive phase variations of waves propagating in fibers to produce intensity variations, resulting in better sensitivities compared to many pure-intensity sensors [3]. A kind of in-fiber interferometer is a multimode interference (MMI) sensor, where interference between the waves traveling in a multimode fiber results in sensor transmission features that are related to the sensed fiber conditions. The intensity patterns resulting from this modal interference depend critically on the geometry and refractive index profile of the used multimode fibers. Many interesting environmental conditions cause perturbations to these fiber properties, for example, bending, mechanical, and thermal stresses. Processing of the obtained interference patterns could be utilized to get information on the mechanical and thermal conditions of the sensing multimode fibers [4]. The use of MMI sensors in different applications, including environmental and structural monitoring, is a subject of current research interest. For static and low-speed applications, such as environmental and static strain monitoring, where spectral measurements can be used, these sensors have the advantage of moderate transmission peak widths, which can be controlled by the length of their sensing multimode fiber sections. This suggests their possible use with low-cost, moderate-resolution spectrometers. Another important advantage is the ease of their use in temporal measurements, in which changes in the sensor transmission at a specified operating wavelength can be used to detect time-varying fiber conditions. This facilitates their use in dynamic structural and machine monitoring as well [5].

MMI sensor structures, principles of operation, and approaches to sensing are reviewed in the next section. Their potential use in different applications and possible sensor multiplexing schemes are discussed next. Challenges facing their industrial sensor realizations are then exposed, and conclusions are drawn at last.

2. Structures and principles of operation

2.1 Sensor structures

The basic structure of a multimode interference fiber sensor consists of a length of a fiber where more than one propagating mode coexist, whether as core- or cladding-guided modes, excited by a lead-in fiber and terminated in a lead-out fiber, which are typically single mode. Interference between the propagating modes in the multimoded fiber section provides the means for sensing, while the lead-in and lead-out fibers act as modal filters to ensure a consistent sensor excitation, independent of the source coupling conditions and to deliver the sensor output in a consistent way, independent of the sensor to detector distance [6].

Traditionally, the multimoded fiber section is a length of a multimode fiber (MMF), with either a step-index or graded-index profile [7], fusion sliced to single mode fibers (SMFs) on both ends, comprising a single mode—multimode—single mode (SMS) structure, as shown in **Figure 1**. The MMF can also be a two-mode fiber,

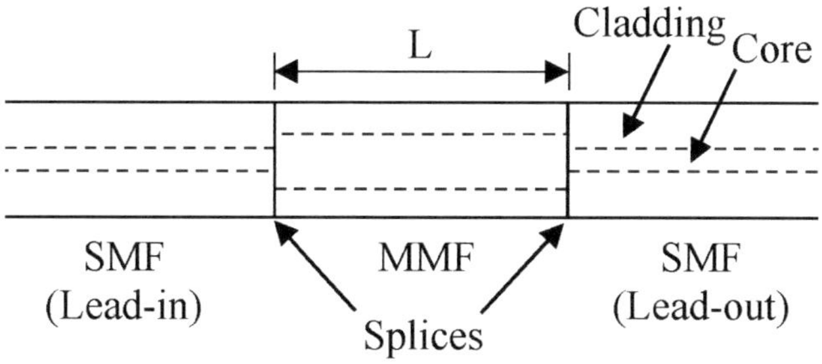

Figure 1.
A schematic of a single mode—multimode—single mode (SMS) fiber structure. Dashed lines separate regions of different refractive indices. L is the length of the multimode fiber section.

a no-core fiber (NCF), or a photonic crystal (PC) one [8]. However, single mode fiber sections where both the core-guided mode and some cladding-guided mode(s) are excited have also been used as multimoded fiber sections. In this case, the sensor fiber structure would have some means to excite both core and cladding modes of the sensing section, for example, by offset splicing the SMF ends or incorporating long-period gratings or multimode fiber sections at its ends [3]. In general, the fewer the modes involved in the modal interference, the simpler the interference spectrum obtained, as will be explained below.

2.2 Principles of operation

Upon excitation of the MMI sensor, the incident field of the lead-in fiber excites modes of the MMF with excitation coefficients that depend on the strength of coupling of each excited mode with the input modal field. As fields of the excited modes propagate with different propagation constants, modal interference occurs along the MMF section. A beating pattern of the field along the multimode section is produced, where constructive interference between two modes with propagation constants β_m and β_n occurs when the following condition holds:

$$(\beta_m - \beta_n)\, z = 2\, N\, \pi \tag{1}$$

where z is the propagation distance along the multimode section and N is an integer. Alternatively, destructive interference between the two modes occurs when the following condition holds:

$$(\beta_m - \beta_n)\, z = (2\, N + 1)\, \pi \tag{2}$$

The total interference field at the multimode section end then couples to the lead-out fiber, according to the strength of coupling of each excited mode with the output fiber modal field. The portion of the input power collected by the output fiber modal field gives the transmission of the fiber structure at the operating wavelength. For perfectly similar lead-in and lead-out connections to the MMF section, the transmission of the structure can be expressed as [9]:

$$T(L, \lambda) = \left|\Sigma_m c_m{}^2 \exp\left(j\beta_m L\right)\right|^2 = \left|\Sigma_n c_n{}^2 \exp[-j(\beta_m - \beta_n)\, L]\right|^2 \tag{3}$$

where c_m and β_m are the excitation coefficient and the propagation constant of the mth mode of the MMF at the wavelength λ, L is the length of the MMF section, and j is the complex root. The transmission spectrum observed versus wavelength and the sensitivity of its features to the fiber mechanical and environmental changes thus depend on the modal characteristics and length of the MMF section. In general, the longer the MMF section, the faster the transmission variation with wavelength [6], as variations in modal propagation constant differences with wavelength are then reflected with larger phase variations in Eq. (3).

In the simplest case of a two-mode interference, Eq. (3) gives a sinusoidal variation of the transmission with the multimode section length, L, and an almost-sinusoidal transmission spectrum versus wavelength [8]. This transmission variation with wavelength, for a given multimode section length, is a result of the difference in modal propagation constants being related to the (free space) wavelength, λ.

The situation gets a bit complicated as the number of modes involved in the interference increases. If the excited modes of the MMF collectively satisfy the condition of constructive interference given by Eq. (1) at a length L_i of the MMF, that is $(\beta_m - \beta_n)L_i = 2\, q_{mn}\, \pi$ and $(\beta_m - \beta_{m+1})L_i = 2\, p\, \pi$, where q_{mn} and p are integers, thus the modal propagation constants, even approximately, satisfy the relation:

$$(\beta_m - \beta_n)/(\beta_m - \beta_{m+1}) = q_{mn}/p \tag{4}$$

then an image of the exciting field is reproduced at the length L_i of the MMF, known as the self-imaging length, given by:

$$L_i\,(\lambda) = p\, 2\,\pi/(\beta_m - \beta_{m+1}) = p\, L_{Bm}(\lambda) \tag{5}$$

where $L_{Bm}(\lambda)$ is the beating length of the mth and $(m+1)^{th}$ modes of the MMF at the operating wavelength. For a symmetric structure with a length L_i of the MMF section, the power coupled to the lead-out fiber at the multimode section end is maximized, and Eq. (3) gives a transmission of approximately unity. Alternatively, for a length L of the MMF, the self-imaging condition of Eq. (5) would be satisfied at a certain operating wavelength, λ_i, at which the length L is the MMF self-imaging length, i. e.

$$L = p\, L_{Bm}(\lambda_i) \tag{6}$$

At the wavelength λ_i, self-imaging takes place in the MMF section of length L, and a maximum occurs in the transmission spectrum of the structure [9]. In-between self-imaging positions, a beating pattern of the transmission spectrum is observed, which depends on the modal characteristics of the MMF used.

For axially symmetric structures with parabolic graded-index MMFs, where symmetric modes have modal propagation constants that are nearly uniformly spaced, all excited (symmetric) modes approximately satisfy Eq. (4) with $p = 1$, and the transmission pattern appears like that of a two-mode interference [4]. Self-imaging peaks appear at the positions given by Eq. (6). Halfway between self-images, interference of the modes results in a minimum field at the aperture of the lead-out SMF. The power coupled to the lead-out fiber is then minimized, and minima of the transmission occur.

For step-index MMFs, a complicated spectral pattern is observed. Excited symmetric modes in a weakly guiding MMF approximately satisfy Eq. (4) as well, but with a different value of the integer p for each modal index m [10]. This gives a number of beating sub-images between self-images, according to the orders of the beating modes, in a cyclic beating pattern. The transmission spectrum then have many sub-image peaks of different transmission values, according to the strength of excitation of involved modes, where constructive interferences between the modes occur. These peaks are separated by transmission troughs, where destructive modal interferences occur.

Thus, modal interference in the MMF section results in an SMS transmission spectrum with peaks at certain wavelengths, λ_N, where some modes satisfy Eq. (1) of constructive interference across the multimode section:

$$(\beta_m - \beta_n)\, L = \Delta\beta_{mn}(\lambda_N)\, L = 2\, N\, \pi \tag{7}$$

and troughs or dips at other wavelengths, $\lambda_{(N\,+\,1/2)}$, where some modes satisfy Eq. (2) of destructive interference across the multimode section:

$$(\beta_m - \beta_n)\, L = \Delta\beta_{mn}\left(\lambda_{(N+1/2)}\right) L = (2\, N + 1)\, \pi \tag{8}$$

These spectral features are sensitive to the fiber mechanical and environmental changes, which could induce changes in relative modal excitation, modal propagation constant differences, and MMF length. In particular, the wavelength locations of spectral peaks or dips would then change to accommodate the induced changes, so that Eq. (7) or Eq. (8) is satisfied at the respective new wavelength values. For example, when the length of a step-index MMF section of an SMS structure is incremented, the structure peak transmission wavelength is found to shift to shorter wavelength values. This behavior was explained by considering the approximate linear dependence of the propagation constant differences on wavelength [9]. Strain and temperature changes, however, affect both propagation constant differences and MMF length, resulting in wavelength shifts of transmission spectral features. These spectral features' shifts can be inverted to give a measure of the applied strain or change in temperature. Alternatively, variation in the structure transmission at a suitable fixed wavelength can be used to trace changes in the MMF conditions. These sensing approaches are explained next.

2.3 Sensing approaches

2.3.1 Spectral measurements

Spectral measurements are carried out to characterize the transmission of an MMI sensor versus wavelength and trace the changes in its spectral features with exposition to the quantity to be sensed. A typical spectral measurement setup is illustrated in **Figure 2**. In this setup, the output of a broadband light source, or a widely tuned laser, is coupled by single-mode fibers to the SMS sensor through an isolator. The light

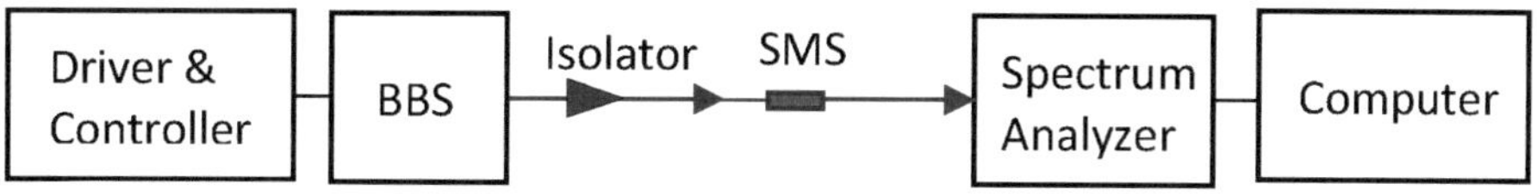

Figure 2.
Spectral measurement setup for MMI sensing. BBS stands for broadband source.

transmitted through the sensor, whose attenuation versus wavelength modulates the transmitted spectrum, is then analyzed by the spectrum analyzer. The transmission of the sensor is obtained by dividing the detected spectrum by the reference source spectrum, which is detected without connecting the SMS sensor.

The spectra obtained of an SMS sensor experiencing a progressively increasing axial strain, in an experiment to monitor the pressure inside an infrastructure pipe [11], are shown in **Figure 3**. The sensor was made of a section of 105-μm core/125-μm cladding diameters, weakly guiding step-index MMF of length about 26.5 mm, spliced on both sides to SMF-28 single mode leads. The pipe was pressurized within its elastic limit so that its wall strain is directly proportion to the applied pressure. The relative detuning of the wavelength of the sensor transmission maximum near 1580 nm, labeled as λ_N in **Figure 3**, versus the strain as measured by a fiber Bragg grating (FBG) mounted in a similar way on the pipe wall, is shown in **Figure 4**. The nearly linear characteristic, with a high R^2 value, of the peak wavelength detuning with increasing strain suggests the feasibility of using the MMI sensor for strain monitoring.

For a weakly guiding step-index MMF excited by a symmetric input field, Eq. (7) can be inverted to give the wavelength λ_N in terms of the fiber parameters [12]:

$$\lambda_N = 16 n_{core} a^2 N / \{(m-n)\,[2(m+n)-1]\,L\} \tag{9}$$

where n_{core} is the refractive index and a is the radius of the multimode fiber core, respectively. The relative detuning of the transmission peak (with the same N, m, and n integer values), with the fiber axial strain, ε, and change in temperature, ΔT, is accordingly given as:

$$\Delta\lambda_N/\lambda_N = -[1 + 2\nu + P_e]\,\varepsilon + [\alpha + \xi]\,\Delta T \tag{10}$$

where ν is Poisson's ratio, P_e is the effective strain-optic coefficient, α is the thermal expansion coefficient and ξ is the thermo-optic coefficient of the fiber material. Eq. (10) indicates a linear detuning of λ_N with axial strain or temperature change of the MMF and explains the linear negative detuning of λ_N with the axial strain, as

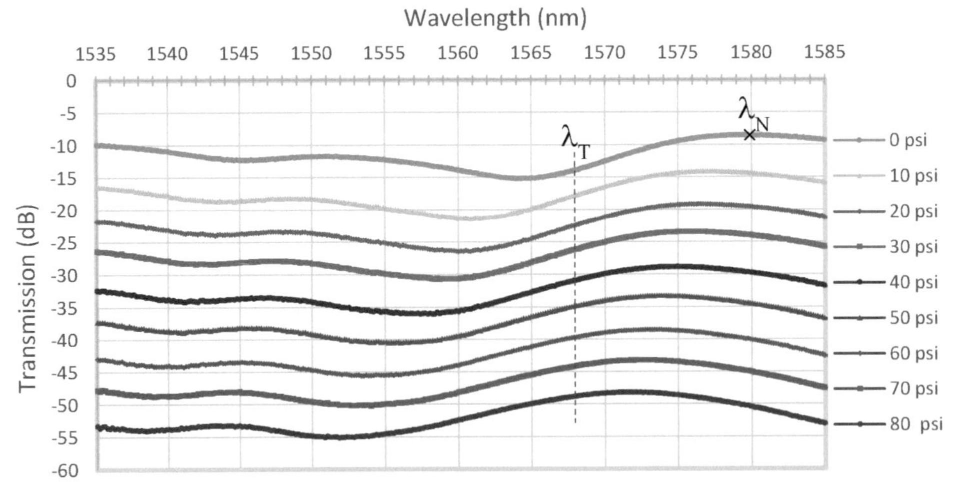

Figure 3.
SMS transmission spectra obtained upon increasing the pressure inside the pipe, shifted by 5 dB from each other for clarity.

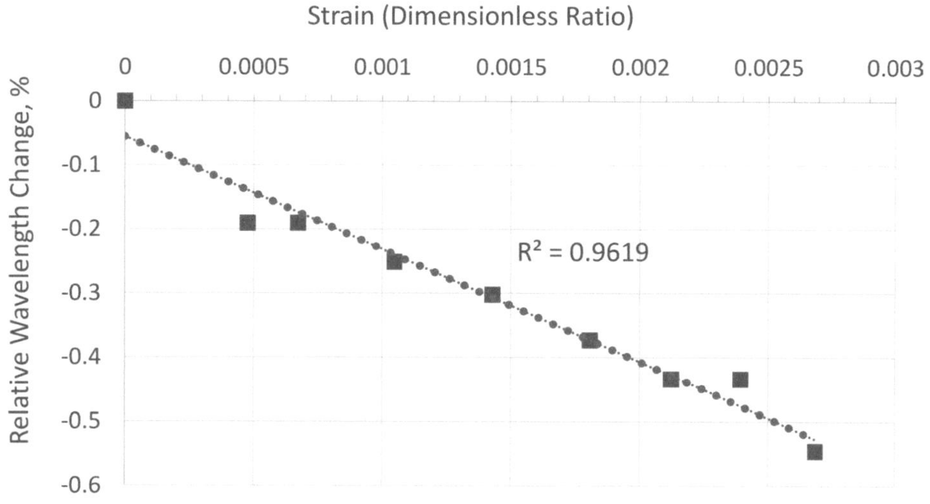

Figure 4.
Relative detuning of the sensor transmission maximum labeled as λ_N in ***Figure 3****, versus the strain as measured by a similarly strained FBG. Squares represent experimental measurement data, whereas the dashed line is a linear fit of the experimental data.*

observed in **Figure 4**. For a Poisson's ratio of 0.17 and a value of P_e around 0.22 of glass fibers, a sensitivity of the relative detuning to strain of −1.56 is obtained, which is close to the value −1.755 obtained from **Figure 4**.

In addition to the method described above, of direct tracing of a transmission spectral feature detuning against the sensed quantity, other spectral measurement approaches have been suggested. Although Eq. (10) can theoretically be obtained for every constructive peak or destructive trough of a transmission spectrum, it is typically found that different features of an MMI transmission spectrum detune slightly differently. This adds ambiguity to the measurement as the selection of the spectral feature to trace is arbitrary. It can be observed from **Figure 3**, however, that the transmission spectra have similar shapes with a definite trend to move to shorter wavelengths with increasing strain. A method making use of this observation to overcome peak selection ambiguity and improve sensor reliability is spectrum differential integration. In this method, differences between spectra, integrated over a range of wavelengths, are used instead of specific feature detuning, to relate to the sensed quantity [13].

Another method of spectral measurement reverts to the Fourier analysis of the sensor transmission spectra [8, 14]. Fourier components, at spatial frequencies corresponding to beating between excited modes, are used to relate to the sensed quantities. These spatial component phases are found to relate with good linearity to the axial strain and temperature changes of fibers. Although this method of measurement needs more computation than direct spectral measurements, it alleviates the dependence of measurements on the MMF length tolerance and allows for possible multiplexing of MMI sensors [8].

A different approach to spectral measurement uses an MMI sensor in an active fiber-loop laser system arrangement. The selective transmission properties of the MMI sensor are then used to determine the laser output wavelength according to the sensed quantity. Such systems give high sensitivity and large signal-to-noise ratio measurements [15].

Figure 5.
Block diagram of an MMI temporal measurement setup. DAQ stands for data acquisition system.

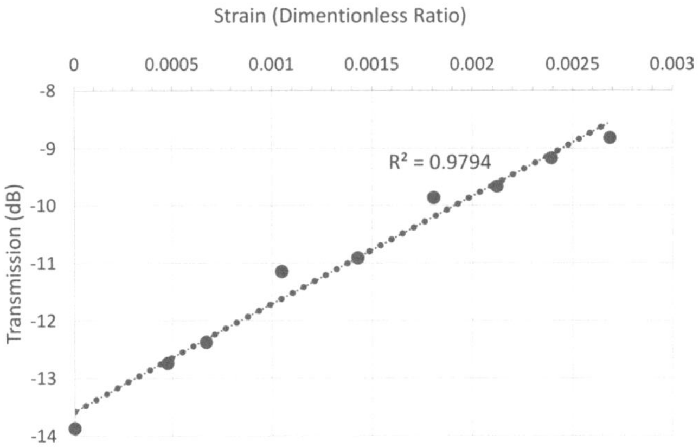

Figure 6.
*Transmission of the SMS of **Figure 3** at the wavelength labeled as λ_T, versus the strain as measured by a similarly strained FBG. Circles represent experimental measurement data, whereas the dashed line is a linear fit of the experimental data.*

2.3.2 Temporal measurements

In this type of measurement, time variation in the sensor transmission at a suitable fixed wavelength is used to trace changes in the sensed quantity with time [5, 16]. This facilitates sensor use in dynamic sensing, such as structural and machine vibration monitoring, where the use of popular sensors, like FBGs, is significantly expensive [17]. **Figure 5** shows a block diagram of an MMI sensor temporal measurement setup. Instead of the broadband light source used for spectral measurement, a narrow band source or a laser is used to excite the sensor at a specific operating wavelength, at which the sensor transmission is evaluated. The variation in the sensor transmission at the operating wavelength then follows changes in the sensor conditions as functions of time.

For illustration, **Figure 6** gives the transmission of the SMS of **Figure 3** at the wavelength labeled in the figure as λ_T versus the exerted strain. Monitoring the transmission of the MMI sensor at this operating wavelength, enables a monitoring of the strain variations induced by pipe vibrations, for example, due to pipe drilling [11]. A linearity of the transmission versus strain characteristics would enable a faithful determination of the vibrations' frequency spectrum, which is a valuable tool for discriminating among pipe vibration events.

3. Potential applications

3.1 Static monitoring

In these applications, changes in the quantity desired to be sensed are relatively slow, allowing for a spectral scanning of milliseconds duration between each measurement readout. MMI sensors have been widely used for this type of measurement. In addition to strain and temperature measurements [11, 18], measurements of other mechanical and physical quantities [19], as well as fiber surrounding refractive index [13], have been reported. Fiber strain and surrounding liquid refractive index measurements have been achieved and optimized using etched polymer and no-core MMF sections [20, 21]. Measurement of the refractive index of fiber surroundings, in particular, enables a lot of chemical [15], environmental [22, 23], and biomedical [24] potential sensor applications.

3.2 Dynamic monitoring

In dynamic applications, changes in the quantity to be sensed are relatively fast, not permitting a spectral scanning measurement. For such applications, temporal measurements, as described in Art. 2.3.2, have been successfully used. Machine vibrations [5], dynamic strain variations [11, 16], and fiber vibrations induced by a fluid flow [25] were detected using MMI sensors. Sensors with longer MMF sections were used as disturbance [10, 26] and acoustic [27] detectors.

3.3 Multiplexing schemes

As an advantage over wired gauges, the networking of optical sensors using optical fibers has been of great industrial interest. Different multiplexing techniques have been explored and evaluated [1, 2]. While a simple wavelength division multiplexing scheme has successfully been used for multiple FBG sensor measurements [28], the multiplexing of MMI sensors is still a challenging research subject [29].

Using direct spectral multiplexing, series [30] as well as parallel [31] connections of MMI sensors, have been investigated. They were found to have noticeable individual sensor spectral feature shifts upon connection. Thus, a calibration of the individual sensors before multiplexing would be useless. In addition to sensors' cross talk, high device losses limit the signal-to-noise ratio in a series connection, while signal splitting and combining losses limit parallel connections. Resorting to Fourier analysis techniques [8], the demodulation of parallel [32] and series [29, 33] multiplexed sensors was reported, still with noticeable sensor cross talks. For temporal sensing measurements, multiplexing of MMI sensors using optical time domain reflectometry (OTDR) was suggested [34]. This mandates an operation of all multiplexed sensors at the same wavelength. The OTDR equipment cost would also contribute significantly to the overall sensing system budget.

4. Challenges for industrial realization

4.1 Sensor reproducibility and need for calibration

The sensitivity of an MMI sensor characteristics to the actual length of the MMF section, which is difficult to control and repeat in different sensors, is a major challenge facing MMI sensor realization. In spite of the possibility of controlling the sensor transmission feature locations and widths by the MMF section length [35], it is difficult to precisely control actual fiber section lengths to meet sensor designs. Therefore, there is a need for individual sensor calibration before installation for measurement, which is a practical difficulty. Exploring methods to extract the sensed variations independent of the individual sensor multimode section length would thus be an important step to promote practical sensor deployment.

4.2 Packaging

Unpackaged, bare fiber sensors are generally very fragile, having lower stability and endurance than packaged ones. A sensor package needs to provide protection, close fiber-to-base thermal expansion match, and good strain transfer to the sensor. Otherwise, bare sensors may be used, which necessitates the manipulation of fragile bare fibers with no possibility of sensor reuse or measuring negative strain (compression). Due to strain-temperature cross-sensitivity, it is further necessary to measure, control, or compensate for undesired cross-sensitivity effects [36]. Packaging solutions already developed for FBG sensors can, however, be adapted to MMI fiber sensors [37].

4.3 Sensors' multiplexing

Sensor multiplexing is an important aspect of optical sensors, which enables multipoint sensing by optical fiber networking. Multiplexed sensors share the same resources, resulting in a reduced cost per sensor as well as simple wiring installations. The realization of MMI sensor networking configurations and demodulation techniques suitable for operating many sensors simultaneously is thus very important in order to promote their industrial deployment. MMI sensor multiplexing schemes reported to date show significant sensor cross-talks, and a quantitative comparison of cross-talk in different networking approaches is still lacking [29]. Exploring the best performance and cost-wise scenarios to operate multiple MMI sensors is thus a research task that still needs to be done.

5. Conclusions

In this chapter, simple optical fiber sensors based on modal interference in multimode optical fibers were addressed. Their sensor structures and principles of operation were introduced. Approaches to their use in sensing were explained and contrasted. Their potential applications and possible sensor multiplexing schemes were briefly reviewed, and challenges facing their industrial realizations were exposed. It is hoped that this review of such an emerging fiber sensing technique will help to promote the achievement of simpler and more affordable optical fiber sensing technology to serve various industrial and civil sectors.

Acknowledgements

Experiments supporting this work were conducted with the assistance and administrative help of Prof. Raghied M. Atta, while the author was on leave to the Electrical Engineering Department, College of Engineering, Taibah University, Almadinah Almunawwarah, K.S.A.

Author details

Ahmed Hisham E. Morshed
Electronics and Electrical Communication Engineering Department, Faculty of Engineering, Ain Shams University, Cairo, Egypt

*Address all correspondence to: ahmed_morshed@eng.asu.edu.eg

References

[1] Culshaw B, Kersey A. Fiber-optic sensing: A historical perspective. Journal of Lightwave Technology. 2008;**26**: 1064-1078. DOI: 10.1109/JLT.0082.921915

[2] Udd E. An overview of fiber optic sensors. Review of Scientific Instruments. 1995;**66**:4015. DOI: 10.1063/1.1145411

[3] Lee BH et al. Interferometric fiber optic sensors. Sensors. 2012;**12**: 2467-2486. DOI: 10.3390/s120302467

[4] Tripathi S, Kumar A, Varshney R, Kumar Y, Marin E, Meunier J-P. Strain and temperature sensing characteristics of single-mode–multimode–single-mode structures. Journal of Lightwave Technology. 2009;**27**:2348-2356. DOI: 10.1109/JLT.2008.2008820

[5] Morshed AH, El-Sayed IM. Monitoring of vibrations using multimode optical fiber sensors. In: Proceedings of the 33rd National Radio Science Conference (NRSC2016), February 22–25. Aswan, Egypt; 2016. pp. 384-389. DOI: 10.1109/NRSC.2016.7450863

[6] Wu Q et al. Singlemode-multimode-Singlemode fiber structures for sensing applications—A review. IEEE Sensors Journal. 2021;**21**:12734-12751. DOI: 10.1109/JSEN.2020.3039912

[7] Kumar M, Kumar A, Tripathi S. A comparison of temperature sensing characteristics of SMS structures using step and graded index multimode fibers. Optics Communications. 2014;**312**: 222-226. DOI: 10.1016/j.optcom.2013.09.034

[8] Barrera D et al. Low-loss photonic crystal fiber interferometers for sensor networks. Journal of Lightwave Technology. 2010;**28**:3542-3547. DOI: 10.1109/JLT.2010.2090861

[9] Wang Q, Farrell G, Yan W. Investigation on single-mode–multimode–single-mode fiber structure. Journal of Lightwave Technology. 2008; **26**:512-519. DOI: 10.1109/JLT.2007.915205

[10] Morshed AH. Multimode optical fiber sensors based on self-imaging. In: 28th National Radio Science Conference (NRSC). Cairo, Egypt; 2011. pp. 1-8. DOI: 10.1109/NRSC.2011.5873638

[11] Morshed AH, Atta RM. Multimode optical fiber strain monitoring for smart infrastructures. Ain Shams Engineering Journal. 2023;**14**:102181. DOI: 10.1016/j.asej.2023.102181

[12] Li E. Sensitivity-enhanced fiber-optic strain sensor based on interference of higher order modes in circular fibers. IEEE Photonics Technology Letters. 2007;**19**:1266-1268. DOI: 10.1109/LPT.2007.902271

[13] Li Y, Harris E, Chen L, Bao X. Application of spectrum differential integration method in an in-line fiber Mach-Zehnder refractive index sensor. Optics Express. 2010;**18**:8135-8143. DOI: 10.1364/OE.18.008135

[14] Xu Y et al. Dispersion effects of high-order-mode fiber on temperature and axial strain discrimination. Photonic Sensor. 2015;**5**:224-234. DOI: 10.1007/s13320-015-0249-9

[15] Lan X et al. Fiber ring laser interrogated zeolite-coated singlemode-multimode-singlemode structure for trace chemical detection. Optics Letters.

2012;**37**:1998-2000. DOI: 10.1364/OL.37.001998

[16] Ran Y et al. Vibration fiber sensors based on SM-NC-SM fiber structure. IEEE Photonics Journal. 2015;**7**:6800607. DOI: 10.1109/JPHOT.2015.2408436

[17] Campanella CE, Cuccovillo A, Campanella C, Yurt A, Passaro VMN. Fibre Bragg grating based strain sensors: Review of technology and applications. Sensors. 2018;**18**:3115. DOI: 10.3390/s18093115

[18] Noor SFSM, Harun SW, Ahmed H, Muhammad AR. Multimode interference based fiber-optic sensor for temperature measurement. Journal of Physics: Conference Series. 2019;**1151**:012023. DOI: 10.1088/1742-6596/1151/1/012023

[19] Guzmán-Sepúlveda JR, Guzmán-Cabrera R, Castillo-Guzmán AA. Optical sensing using fiber-optic multimode interference devices: A review of nonconventional sensing schemes. Sensors. 2021;**21**:1862. DOI: 10.3390/s21051862

[20] Nakanishi T et al. Multimode interference-based strain sensing using micro dry-etched perfluorinated polymer optical fibers. Japanese Journal of Applied Physics. 2024;**63**:028003. DOI: 10.35848/1347-4065/ad1e9e

[21] Li Y, Liu Z, Jian S. Multimode interference refractive index sensor based on coreless fiber. Photonic Sensor. 2014;**4**:21-27. DOI: 10.1007/s13320-013-0137-0

[22] Joe HE, Yun H, Jo SH, Jun MBG, Min BK. A review on optical fiber sensors for environmental monitoring. International Journal of Precision Engineering and Manufacturing-Green Technology. 2018;**5**:173-191. DOI: 10.1007/s40684-018-0017-6

[23] Zain MFIM, Husein NAB. Multimode interference self-imaging optical fiber sensor based on sol-gel silica for methane detection. Journal of Physics: Conference Series. 2023;**2432**:012014. DOI: 10.1088/1742-6596/2432/1/012014

[24] Socorro AB et al. Fiber-optic Immunosensor based on an etched SMS structure. IEEE Journal of Selected Topics in Quantum Electronics. 2017). Art no. 5601808;**23**:314-321. DOI: 10.1109/JSTQE.2016.2633819

[25] Morshed AH, Atta RM, Packirisamy M. Fluidic flow measurement using single mode–multimode–single mode optical fiber sensor. IEEE Sensors Journal. 2021;**21**:13316-13326. DOI: 10.1109/JSEN.2021.3069363

[26] Ribeiro RM, Balod YC. Modalmetric interferometer under fingerprint disturbances: The effect of lateral offset between the singlemode and multimode fiber splice. Microwave and Optical Technology Letters. 2017;**60**:151-157. DOI: 10.1002/mop.30932

[27] Freitas TAMG, Silva VH, Ribeiro RM. Sensitivity of fiber-based modalmetric devices intended for optical detection of acoustic signals. Microwave and Optical Technology Letters. 2019;**62**:999-1008. DOI: 10.1002/mop.32136

[28] Chen J, Liu B, Zhang H. Review of fiber Bragg grating sensor technology. Frontiers of Optoelectronics China. 2011;**4**:204-212. DOI: 10.1007/s12200-011-0130-4

[29] Liu Q, Xia Y, Xue P, Cai L, Wu Q, Fu Y. Research on photonic crystal fiber sensor network based on fast Fourier

transform Bandpass filtering. IEEE Transactions on Instrumentation and Measurement. 2023). Art no. 9507608; **72**:1-8. DOI: 10.1109/TIM.2023.3276005

[30] Liu X, Zhang X, Liu Y, Liu Z, Peng W. Multi-point fiber-optic refractive index sensor by using coreless fibers. Optics Communications. 2016; **365**:168-172. DOI: 10.1016/j.optcom.2015.10.071

[31] Fuentes-rubio Y-A, Domínguez-Cruz R-F, Guzmán-Sepúlveda J-R. Multipoint fiber optics refractive index sensor based on multimode interference effects. Applied Optics. 2021;**60**: 9691-9695. DOI: 10.1364/AO.439151

[32] Liu L, Gong Y, Wu Y, Zhao T, Wu H-J, Rao Y-J. Spatial frequency multiplexing of fiber-optic interferometric refractive index sensors based on graded-index multimode fibers. Sensors. 2012;**12**:12377-12385. DOI: 10.3390/s120912377

[33] Galarza M, Perez-Herrera RA, Leandro D, Judez A, López-Amo M. Spatial-frequency multiplexing of high-sensitivity liquid level sensors based on multimode interference micro-fibers. Sensors and Actuators A: Physical. 2020; **307**:111985. DOI: 10.1016/j.sna.2020.111985

[34] Hatta AM, Indriawati K, Bestariyan T, Humada T, Sekartedjo. SMS fiber structure for temperature measurement using an OTDR. Photonic Sensors. 2013;**3**:262-266. DOI: 10.1007/s13320-013-0104-9

[35] Wang K et al. Fiber-Optic Multimode Interference Sensing: Comprehensive Characterization and its Potential for Strain-Insensitive Temperature Sensing. Ithaca, NY: Cornell University; 2022. DOI: 10.48550/arXiv.2204.11044

[36] Wang K et al. Advances in optical fiber sensors based on multimode interference (MMI): A review. IEEE Sensors Journal. 2020;**21**:132-142. DOI: 10.1109/JSEN.2020.3015086

[37] Lin YB, Chang KC, Chern JC, Wang LA. Packaging methods of fiber-Bragg grating sensors in civil structure applications. IEEE Sensors Journal. 2005; **5**:419-424. DOI: 10.1109/JSEN.2005.844539

Chapter 5

Spatial Multiplexing on a Multicore Fiber: A Prospective Technology for Fiber-Optic Communication Links to Reach a Petabyte Capacity

Mikhail E. Belkin, Sergei L. Semjonov, Olga N. Egorova, Eugeniy Plastinin, Mikhail G. Vasil'ev and Alexander S. Sigov

Abstract

The chapter first provides an analytical review of scientific literature justifying the possibility, desirability, and effectiveness of the additional use of spatial multiplexing technology in the design of fiber-optic communication systems in such areas that are rapidly developing in the world: fifth generation cellular networks in a megapolis and interconnecting links of hyper-scale data centers. Then, the optimal design principles are considered, and the basic requirements for the parameters and characteristics of the used components and units of the transmission and receiving equipment are given. This article describes the results of developing a conceptual schematic diagram for an interconnect communication link between hyper-scale data centers, including terminal transceivers using time and spectral multiplexing and demultiplexing and an optical path with spatial multiplexing and demultiplexing based on a multicore optical fiber and a multicore optical connector. The results of developing a conceptual schematic diagram of an interconnection communication link for hyper-scale data centers with record key parameters: throughput of more than 4 Tbit/s with specific energy consumption of no more than 0.1 nJ/bit are described. To verify the proposed scheme, the results of computer modeling, prototyping, and experimental studies of a fiber-optic path based on a 7-core optical fiber are presented and discussed.

Keywords: ultra-high-speed fiber-optic digital communication system, hyper-scale data center, interconnect communication link with time and spectral multiplexing and demultiplexing, optical path with spatial multiplexing and demultiplexing, multicore optical fiber

1. Introduction

As is known, the total traffic of the worldwide digital fiber-optic communication systems, the most important and significant implementations of which are fifth-generation cellular telecommunication networks (5G NR) in a megapolis [1] and hyper-scale data center (DC) interconnection links (ICL) [2], has increased dramatically over the past decade. In particular, the main reason for these areas is the ever-growing popularity of modern Internet applications, such as cloud computing, streaming video, and social networks. For example, according to Cisco statistics, DC traffic will continue to grow at an annual growth rate of 25%, reaching more than 20 zettabytes last year. To ensure and support this traffic, hyper-scale DCs have been built and are operating in almost all developed countries, the effective networking of which requires the use of ultra-high-speed, interference-protected ICLs. The most suitable carrier for solving this problem is undoubtedly optical fiber, which is confirmed by the long-term development of digital fiber-optic transmitting systems (FOTS), which currently provide the main traffic in both transport and local communication systems.

It is obvious that the studied ICLs, both between 5G's central and base stations and between several hyper-scale DCs, in terms of design principles fully correspond to ultra-high-speed fiber-optic links of digital FOTS, the main units of which are active optical transceivers (OTC) implementing direct and reverse electro-optical conversions and a passive fiber-optic path (FOP) connecting them. Additionally, an important distinguishing feature in terms of efficient operation of multicore FOP for DCs is the strict limitation of energy consumption, which is associated with the general problems of modern DC development.

The novelty of the proposed solution lies:

a. At the system level:

A newer design principle based on the standard use of spectral separation of channels in OTCs with the addition of passive spatial separation of channels in the fiber-optic cable connecting them allows us to significantly increase the overall throughput of the FOP without increasing the energy costs for its operation.

b. At the device level:

In the OTC design, instead of the standard semiconductor laser with distributed feedback (DFB), which does not provide the newer requirement for specific energy consumption, a vertical-cavity surface-emitting laser (VCSEL), the power consumption and cost of which are an order of magnitude lower. Besides, to meet the increased time-division multiplexing requirements for overall throughput, instead of the standardly adopted pulse amplitude modulation (PAM), the optical injection locking (OIL) method of the VCSEL under study, which in addition to expanding the band of its direct modulation leads to an increase in the length of the ICL due to the weakening of the effect of optical fiber dispersion.

Following it, this chapter addresses a prospective technology for fiber-optic communication links to reach up to a petabyte capacity due to additional employment of spatial multiplexing. In particular, Section 2 reviews key applications and state-of-the-art development of digital fiber-optic communications. In addition, Section 3 presents the design principles of prospective fiber-optic telecom systems for 5G NR mobile communication systems and optical interconnects for hyper-scale DCs. The basic composition, key design, and parameters, approaches to reduce crosstalk in multicore optical fibers (MCOF), examples of modern worldwide MCOF production, as well as simulation of the proposed MCOF and experimental results of prototyping two 7-core fiber-optic links are demonstrated and discussed in Section 4. Finally, Section 5 concludes the chapter.

2. Key applications and state-of-the-art development of digital fiber-optic transmitting systems

As is known, there are two options for exchanging information over relatively long distances: over open space and over an artificial guide medium [3, 4]. Until the 1970s, only metallic carriers (two-wire lines, coaxial cables, and round waveguides) were used as the latter. In the first half of the 1970s, after Corning, USA [5] managed to ensure a link attenuation of less than 20 dB/km, the practical development of a new direction began using fiber optics as a transmission medium. Currently, telecommunication FOTS has taken a leading position among various means of communication [6–10]. The main advantages of fiber optics: low losses and dispersion (i.e., broadband), excellent weight, size, and cost characteristics, and insensitivity to electromagnetic interference—have been implemented in practice, and on the basis of fiber optics, back in the last century, trunk land links of digital communication with a length of several tens of thousands of kilometers, as well as transoceanic submarine links connecting all continents, were built. This relatively short historical period has already shown the undeniable advantages of using fiber optics in the construction of transport networks of great length and capacity. Statistical data on the modern development of communication facilities show that with the number of equivalent telephone channels over 10,000, the use of fiber optics is more economical than radio relay and satellite communication systems. The share of digital fiber optics in the field of long-distance communications currently accounts for more than 80% of transmitted channels.

The modern development of digital telecom FOTS intended for the transmission of telephone, fax, television, multimedia, and other signals is moving toward increasing the transmission channel capacity, which, as in telecommunications (radio communications, multichannel wired communications), is achieved through time multiplexing (TMP) [9]. This trend is expressed in increasing the linear speed, and today digital FOTS are produced with a transmission speed of up to 800 Gbit/s over a single optical channel [11]. Further increase in capacity in this case is constrained by the speed of logical devices required for the signal regeneration. The second trend in FOTS is implemented in the form of the so-called spectral multiplexing (SMP) [12]. The third relatively recently appeared approach, called spatial multiplexing (SpMP), is associated with the use of multicore optical fiber (MCOF) [13]. Let us briefly consider the modern principles of implementing these ways of increasing the capacity of digital FOTS to speeds up to in the petabyte range.

3. Design principles of prospective telecommunication systems

According to the accepted worldwide classification, digital FOTS is an ordered set of technical means for transmitting digital information (data) via fiber-optic cable, carried out on one or several optical carriers using time division of channels in the video range and/or spectral division of channels in the optical range. From the very beginning of their development, digital FOTSs were aimed at conquering one of the largest and most promising world markets: the telecommunication market. To do this, first of all, it was necessary to ensure reliable joint operation in field conditions of all three components of the FOTS's skeleton diagram: the optical transmitting module (OTM), the FOP, and the optical receiving module (ORM). Taking into account the features of a fundamentally new information transmission system, research and development works were carried out in parallel in three directions:

- Research and development of an optimal carrier of optical signals, for which a multimode or single-mode silica lightguide was selected, and effective methods for forming FOP laid in the ground, in water, on overhead supports or in telephone ducts;

- Research and development of an optimal emitter (electro-optical converter) of the near infrared (IR) range (from 0.8 to 1.6 μm), for which a semiconductor laser was selected;

- Research and development of an optical receiver (optical-electrical converter) of the near IR range, for which a semiconductor pin-photodiode was selected.

The results of R&D were implemented in the equipment, on the basis of which experimental and commercial communication links were built. The overwhelming majority of digital FOTSs in the 1970s were operated in the so-called first transparency window (in the region of 0.85 μm) on a multimode fiber with a step or gradient profile of the refractive index [9]. The following typical parameters were obtained: attenuation from 2 to 3 dB/km, transmission speed up to 140 Mbit/s, which immediately allowed them to be implemented, for example, in city telephone networks for the organization of multichannel digital ICLs between automatic telephone exchanges. By the beginning of the 1980s, the experiments obtained in laboratories were confirmed in real networks, as a result of which most national administrations created long-term programs for the development of fiber-optic communications. The priority areas were the implementation of FOTS on long-distance telephone networks, the creation of fully fiber-optic national and transnational communication networks, and the connection of continents using underwater FOTS to create a worldwide global communication network.

To implement these ambitious projects, researchers and developers had to make a revolutionary leap in the development of the technology and component base of FOTS. First of all, the need for real-time transmission of orders of magnitude larger information volumes forced the abandonment of the already mastered multimode optical fibers and the transition to single-mode ones with a much wider bandwidth. This step led to a sharp aggravation of the problem of connecting effectively a laser emitter with a fiber and fibers with each other, which had to be implemented with submicron accuracy and high reproducibility. Another need for transmitting optical signals over very long distances forced work in the second and third transparency windows (near 1.3 and 1.55 μm) of silica-based fibers with theoretical values of attenuation of 0.27 and 0.16 dB/km, respectively. As a result, it was necessary to create new semiconductor emitters and receivers based on InGaAs and InGaAsP solid solutions and ensure their reliable and efficient operation in the modulating frequency band of up to several gigahertz [10]. Eventually, by the end of the 1980s, digital FOTSs of the so-called plesiochronous hierarchy with speeds of up to 1 Gbit/s and synchronous hierarchy with speeds of up to 2.5 Gbit/s were created and implemented in networks. The next global task set for fiber-optic technology was to conquer a larger market of local communication networks and subscriber access networks (figuratively called the "last mile"). It should be noted that these networks differ significantly from the already mastered transport networks both in architecture and in the information content transmitted. If the basis of the transport network is cable trunk links connecting two points with an insignificant number of taps, then the characteristic features of local networks (LAN, MAN, etc.) and subscriber access networks consist of branching with a large number of taps and the required transmission speed of up to several tens of Gbit/s [9, 10]. The need for such speeds arose due to the need to deliver to many network subscribers not only voice signals and data, but also video information (for example, television programs), the band of which is more than three orders of magnitude wider than the speech signal.

For a more concrete understanding, the level and directions of modern development of FOTS, two significant examples will be shortly reviewed below: the cellular telecommunications system of the fifth generation in a megapolis and optical ICLs of hyper-scale DCs.

3.1 5G NR mobile communication systems in a megapolis

The ever-growing demands of worldwide science, technology, and business societies for ultra-wideband telecommunications services requiring super-high capacity communication channels have led to the need for a radical improvement in the design principles and key parameters of the developing fifth-generation cellular communication system (5G NR) in comparison with the existing systems of fourth generation (4G LTE). The development of the design principles goes along the way of increasing the efficiency of the existing frequency spectrum, expanding it by introducing new frequency ranges, and enhancing the density of the service area using new and improved network architectures, information carriers, and methods of transmitting digital signals. As a result, it is planned to radically improve the key parameters, the level of which follows from **Table 1** [14–16].

Following the trends described above, an advanced approach to designing a megapolis cellular telecommunications system is needed. Its general essence lies in a thorough and more detailed structuring of the cellular network based on the concept of small cells, containing a sequential partitioning into macrocells, microcells, and picocells [16] with the widespread use of advanced fiber-optic communication methods and approaches at all its structural levels.

To make the analysis more concrete, below we will briefly consider one of the possible examples of designing a 5G NR cellular telecommunications network in the city of Moscow [17]. The distinctive features of the principles of designing the network under study consist of increasing the capacity of the above digital channels with time division, due to the spectral multiplexing in the C- and L-spectral passbands of silica fiber, which is widely used in the modern FOTS, as well as spatial multiplexing based on MCOF, which is only being mastered by the telecommunications industry.

No	Network parameter	4G LTE	5G NR	Gain (times)
1	Specific connection density (per sq. km)	200 K	1 M	5
2	Maximum speed of a mobile user (km/h)	80	500	6
3	Peak data rate in a cell (Gbps)	2	20	10
4	Peak data rate to the network user (Gbps)	0,1	1	10
5	The volume density of network traffic, Tbit/s/sq.km	1	20	20
6	End-to-end latency (ms)	10	1	10
Total gain				600,000

Table 1.
Predicted radical improvement in 5th generation network parameters.

The modern administrative-territorial division of Moscow city includes 12 administrative districts (AD), each of which can be used as a macrocell. Among them, one of the largest is the Western AD (WAD), which includes 1,392,993 inhabitants, located on an area of 153,034.3 square kilometers, with a population density of 9102.49 people per square kilometer. The infrastructure of this AD includes 13 municipal districts (MD), each of which, according to the developed concept, should be controlled by the corresponding microcell base station (BS). General characteristics of all WAD's MDs are listed in **Table 2**.

The diagram of the WAD subnetwork of the 5G NR communication network, built on the basis of the Moscow plan, is shown in **Figure 1**. Note that the digital designations in **Figure 1** correspond to the numbering of the MDs in **Table 2**.

As follows from the topology of this AD, to connect all microcell BSs via optical ICLs, a ring configuration with taps can be used. Namely, the ring circuit is formed along the peripheral MDs (2–10) with taps to the central one (Fili-Davydkovo (1)). For reliability, communication with each microcell BS is duplicated from the neighboring one. The total length of the network is about 100 km.

To assess the capacity of the ICLs in the WAD macrocell, we determined its throughput based on the data in **Table 2**. As follows, the population of WAD is approximately equal to 1 million 393 thousand citizens. Taking into account business users and further development, we took the total number of network users equal to 1.5 million. Based on this, we calculated the average number of users in WAD's MD, taking into account the maximum and minimum number of users, which are located on the territory of MD "Kuntsevo" and MD "Vnukovo," respectively. Using the appropriate factor for each optical ICL connecting the corresponding MDs, we determined the parameters of the digital optical channel, including the transmission rate on one optical carrier and the number of carriers. The calculation results are shown in **Table 3**.

Table 3 clearly shows that to build a 5G NR cellular telecommunications network in a modern megapolis, it is necessary to use a combination of time multiplexing with

No	Name of a MD	Citizen population	Area, sq.km	Population density, per sq.km
1	Fili-Davydkovo	115,128	6.96	16,541.38
2	Filevsky park	94,323	9.62	9804.89
3	Dorogomilovo	76,093	7.95	9571.45
4	Ramenki	141,610	18.54	7638.08
5	Prospekt Vernadskogo	63,659	4.65	13,690.11
6	Troparyovo-Nikulino	124,167	11,27	11,017.48
7	Ochakovo-Matveyevskoye	130,742	17.54	7453.93
8	Mozhaysky	138,610	10.73	12,917.99
9	Kuntsevo	152,364	16.56	9200.72
10	Krylatskoye	82,817	12.04	6878.49
11	Novo-Peredelkino	121,553	8.48	14,334.08
12	Solntsevo	126,456	11.29	11,200.71
13	Vnukovo	25,471	17.42	1462.17

Table 2.
The municipal districts of western AD in Moscow city.

Figure 1.
Diagram of the 5G NR cellular telecommunication system in western AD of Moscow city [17].

a standard digital signal transmission rate of 40 Gbps, spectral multiplexing with a number of optical carriers from 3 to 7, which can be done without any problems using the data of Recommendation ITU-T G.692, and spatial multiplexing based on MCOF. Note that 7 cores are enough to achieve a total network capacity of almost 17 Tbps.

3.2 Interconnect networks for modern data centers

By definition, a DC is a specialized facility that is a connected system of information technology (IT) and engineering infrastructures, the server, and network equipment, which is located in special buildings or rooms connected to external networks, both engineering and telecommunications. In general, it performs the functions of processing, storing and distributing information and is focused on solving business problems by providing information services [18]. The widespread development of DCs began at the end of the last century and is mainly associated with the commercialization of the Internet when fast connections to Internet networks were massively required to ensure a wide presence on the global network. At the same time, the rapid growth in the number of Internet users, as well as the need to increase the volume and quality of communication, have led to a demand for consolidation of existing DCs. The growing need for hardware power led to an increase in energy consumption costs, which even then forced DC's owners to work on optimizing power supply systems.

The emergence of IT giants in the twenty-first century, developing their own IT infrastructures, has led to the formation of so-called hyper-scale DCs with an energy consumption capacity of more than 100 MW, the purpose of which is to provide computing power for leading global players, for example, cloud providers and social networks. This class of DCs must have a developed network of branches connected by an ultra-high-speed ICL and the ability to scale over a wide range. In recent years, the world has seen a significant development of the network architecture of a hyper-scale

No	Name of the receiving μ-cell BS	Optical signal		Optical ICL	
		Speed (Gbps)	Number of carriers	Number of cores	Capacity (Tbps)
1	Fili-Davydkovo	40	5	7	1,4
2	Filevsky park	40	4	7	1,12
3	Dorogomilovo	40	3	7	0,84
4	Ramenki	40	7	7	1,96
5	Prospekt Vernadskogo	40	3	7	0,84
6	Troparyovo-Nikulino	40	6	6	1,44
7	Ochakovo-Matveyevskoye	40	7	7	1,6
8	Mozhaysky	50	5	7	1,75
9	Kuntsevo	50	6	7	2,1
10	Krylatskoye	40	3	7	0,84
11	Novo-Peredelkino	50	4	7	1,4
12	Solntsevo	50	4	7	1,4
13	Vnukovo	40	5	1	0,2
Total		—	62	—	16,89

Table 3.
Parameters of WAD's macrocell ICLs.

DC, which currently contains up to five internal and external levels connected via ICLs that are designed on the basis of two terminal optical transceivers (OTC) and a fiber-optic link (FOL) connecting them with a length of several meters to several kilometers and a standardized throughput of OTCs from 25 to 400 Gbps. However, the modern development of DCs dictates an increase of at least an order of magnitude while maintaining or minimally increasing power consumption, which requires a change in technologies and interface standards [19].

The sizes and growth rate of annual traffic of global DCs [18] at the end of the last decade are illustrated in **Figure 2**. Providing such huge volumes of data was carried out through the development of DC infrastructure, which consisted of both expanding existing and opening new large DCs.

The classification of modern DCs according to the purpose of the occupied space and, accordingly, power consumption [20] is presented in **Table 4**.

ABB statistics showing the evolution of the size and energy load of the largest DCs are illustrated in **Table 5** [21, 22], which shows three periods of growth. If the first period (from 1990 to 2000) was characterized by the presence of a DC with a capacity of up to 2500 racks and a power per rack of about 2 kW, then in the last analyzed period (from 2010 to 2020), it was a DC of 10,000 racks with an average power per rack ranging from 5 to 17 kW. That is, the last three decades of DC development have led to a fourfold increase in the volume of its hardware and an average twentyfold increase in energy consumption.

A typical hyper-scale DC network built on the basis of the abovementioned ICLs, which has tree-type architectures widely used in FOTS (in this case, known as "leaf-spine"), contains five structural tiers [23, 24]. Tiers 1 to 3 form the internal network, and at least tier 5 forms the external network. Moving along the tiers from 1 to 5, the range of

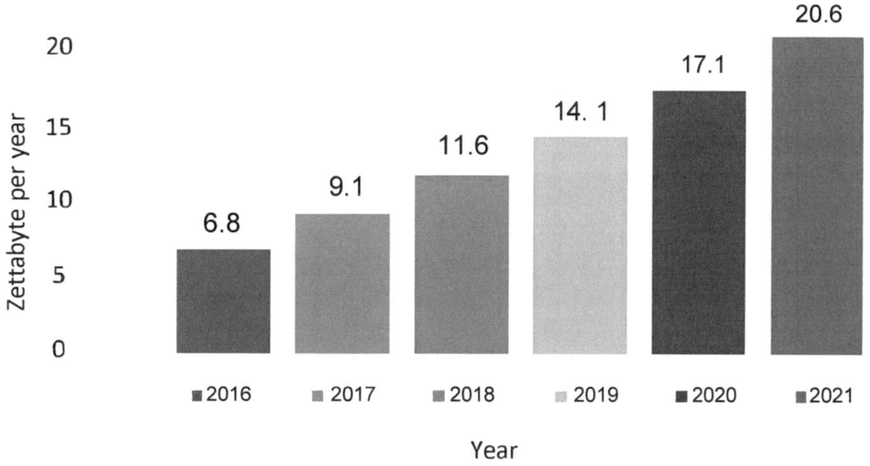

Figure 2.
The volume and growth rate of annual traffic of global DCs [16].

Type	Purpose	Square, sq. m	Power consumption, MW
Micro-DC	Processing Internet of Thing data close to the source	for 1 rack	less than 0.01
Low-scale DC	Enterprise DC for mission-critical data, production data, and more.	up to 500	up to 1
Average-scale commercial DC	Providing DC space to multiple tenants (including remotely managed services)	up to 10,000	up to 10
Big-scale commercial DC	Providing DC space to multiple tenants: individual companies and/or network service providers for the digital ecosystem	up to 50,000	up to 50
Hyper-scale DC	Computing power for leading global players (e.g. cloud providers and social networks). Sufficient capacity for on-demand scalability.	up to 100,000	more than 100

Table 4.
The classification of modern DCs [18].

Area of machine rooms, sq. m	1500	15,000	30,000
Number of racks	2500	5000	10,000
Power on one rack, kW	2	4-8	5-17
Year 1990 2000 2010 2020			

Table 5.
Evolution of the size and energy load of the largest DCs over different periods [21, 22].

each ICL increases from several meters to several kilometers, and the speed increases from 25 to 400 Gbps, which requires changes in technologies and interface standards.

Tier 1. At this lowest tier, individual server racks connect to switches at the top of a cabinet (TOR). Modern DCs typically deploy networks with 25 Gbps throughput, with some artificial intelligence (AI) applications using 50 Gbps speeds. Over the next few years, it is predicted that connection technology will be used with speeds of up to 100, 200, and 400 Gbps. The distances between connections are short, either within a cabinet or to adjacent cabinets, and are typically less than 5 meters. The typical transmission medium used today is copper cable. As speeds advance to 400 Gbps and 800 Gbps, the range of passive cable will be too short, and active electrical cable will be used instead.

Tier 2. At this tier, the first-level switches are connected to the second-level switches (Leaf) using ICLs. The length of such links reaches 50 meters when using the technology of interconnection with speeds of 100 Gbps now and in the future—with speeds of 200 and 400 Gbps. For transmission at the first two speeds, optical transceivers of the 100GBASE-SR4 and 200GBASE-SR4 models operating in the NRZ format are standardly used in combination with multimode fiber. To achieve a speed of 400 Gbps, instead of digital modulation in the NRZ format, it is planned to use PAM4 modulation and optical transceivers of the 400G-DR4, 400GBASE-SR8 models.

Tier 3. The connection of the Leaf and Spine level switches can be located within the main site or external site of the DC with a distance of up to 500 meters. Data transmission occurs at the same speeds and using the same technologies as at the previous tier. But with the increase in the length of the ICL, the transmission medium becomes single-mode fiber and often several parallel operating ICLs using optical transceivers such as 100G-PSM4, 100G-CDWM4, 200GBASE-DR4, and 400GBASE-DR4.

Tier 4. Connection of "spine" and "core" switches can be used for duplex communication between the main and external sites of the DC or two separate DCs with a distance of up to 2 km. As the length of the ICL increases, the use of a large number of parallel optical links will negatively affect the cost of the designed DCs. The solution to this problem is currently the use of SMP technology, which allows data to be transmitted over a single optical fiber at significantly higher speeds. For this approach, OTCs such as 100GBASE-LR4, 100G-CWDM4, and 400GBASE-ER4/−LR4/-FR4 are used.

Tier 5. Typically, this is a connection between two or more DCs of one operator for data exchange, load balancing, or backup after disaster recovery. The length of the ICL can vary from tens of kilometers to hundreds of kilometers. For efficient operation over this fairly long distance, dense wavelength division multiplexing (DWDM) is additionally used, and more recently, coherent fiber-optic communication [25], which has been used in transport FOTS for over 20 years.

3.3 Conceptual block diagram of a telecom ICL based on time, spectral and spatial multiplexing

A conceptual block diagram that interactively links two communication terminals (CT), which can equally well be either cellular base stations in a megapolis (see Section 3.1) or hyper-scale DC (see Section 3.2), is shown in **Figure 3**. According to **Figure 3**, in each CT, the operations of time multiplexing (TMP) and spectral multiplexing (SMP) are sequentially performed. In the FOP, the operation of spatial multiplexing (SpMP), the combined use of which fundamentally makes it possible to implement an ICL with a total capacity of 10 and more than what is currently achieved in megapolis 5G NR or hyper-scale DC networks. The implementation of the TMP is carried out using an L-channel digital modem unit, the task of which is to separate or

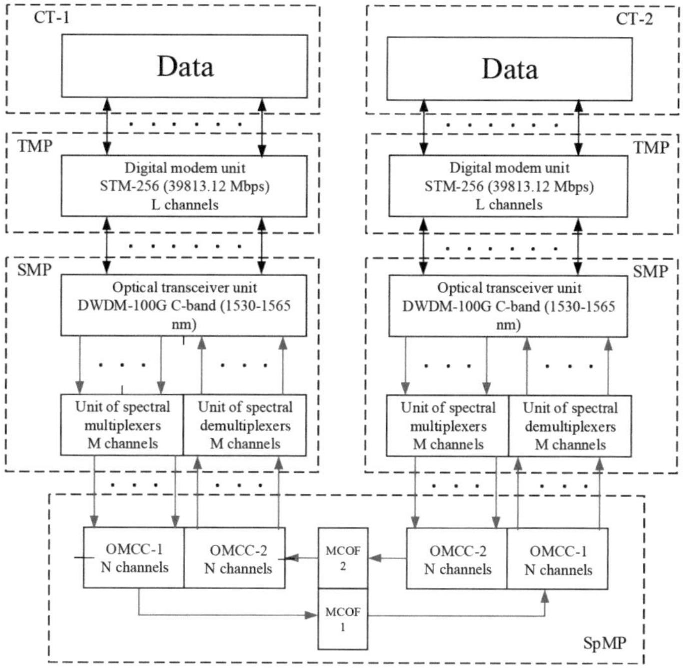

Figure 3.
Proposed conceptual block diagram of two communication terminals interconnected by the ICL [26]. (Black lines are electrical connections, red lines are optical connections).

combine digital electrical channels. In practice, this unit can be excluded if communication is carried out not through a single input/output port, but directly with its individual hardware racks. The subsequent SMP operation is performed using an M-channel unit of OTCs, containing a set of optical transmitting and photoreceiving devices of the same design based, respectively, on semiconductor laser emitters (SLE) and semiconductor photodetectors (SPD). The frequencies of the optical carriers for the SLE must correspond to the standard DWDM grid in the C-band with a step of 100 GHz [12, 27]. To combine and separate modulated optical signals, M-channel units of spectral multiplexers and demultiplexers are used. The optical transceivers of each DC are connected via a 2-fiber FOP containing N-core multicore optical fibers [28] (MCOF-1 and MCOF-2) terminated with N-pin optical multicore connectors (OMCC-1 and OMCC-2).

In the proposed block diagram, low specific energy consumption is ensured due to:

- Relatively low level of TMP since with an increase in transmission speed, the power consumption enhances sharply, which is critical for use in DCs (according to 3.1, no more than 40 Gbps);

- Use of vertically channel surface-emitting lasers (VCSEL) in SMP mode (according to 3.1, 6–8 channels are enough), the power consumption of which is an order of magnitude lower compared to the typically used distributed feedback (DFB) edge emitting laser [11];

- Use of passive ICL based on single-mode 7-core MCOF [26] (according to 3.1, 5–7 cores are enough).

A detailed examination of the proposed scheme (see **Figure 3**) showed that its implementation requires solving a number of serious problems, the main one of which is the development of a fiber-optic path based on a multicore fiber with parameters not inferior to a single-core analog. The results of its solution will be described in the next section.

4. Fiber optic path based on multicore fiber

As noted in the previous section, the way to further increase the capacity of the FOTS is to develop a method of spatial division of channels based on multicore optical fiber (MCOF). In principle, MCOFs were proposed as early as 1979 [29, 30], but at that time this topic did not receive serious development. The interest in MCOF in the last 10 years is primarily associated with the development of FOTS, since the presence of several cores allows for a considerable increase in the data transfer rate over a single fiber. According to estimates, conventional single-core optical fibers (SCOF) allow for the transmission of information with a maximum throughput of about 100 Tbit/s when using multilevel modulation [31]. By using MCOFs, the data transfer rate over a single fiber can be increased by a corresponding number of times. In this case, the communication link throughput will be equal to the product of the throughput of a single core and the number of cores if adjacent cores could be considered as independent communication channels. Using a 12-core MCOF in 2012, the data transfer rate over a single fiber exceeded 1 Pbps [32], and in 2015, another record of 2.15 Pbps was set using a 22-core MCOF [33]. Based on estimates, due to the reduction in the required number of cables, MCOFs are the most promising for use in local FOTSs [28, 30]. In addition, MCOFs are very promising for applications related to data center interconnection links and computer networks since they allow for simplification and reduction in the dimensions of signal transmission paths [34].

Unlike the two abovementioned design principles, this approach has not yet found wide practical application. Nevertheless, the possibility of significantly increasing the throughput, simplifying the design, and decreasing the power consumption of FOTSs has caused high activity in research in this area among world leaders in the production of optical fibers (OF) and the creation of products using them, for example, OFS, Corning, Alcatel-Lucent, NTT Corporation, NEC Corporation, Fujikura, Sumitomo, and their research laboratories and centers. As a result, over the past 10 years, at least several hundred publications have appeared in the scientific literature on the topic of "Multicore optical fibers," devoted to both various design options for such OFs and the creation of their experimental samples, as well as various aspects of their application in FOTS.

4.1 Basic composition, key design, and parameters

Today, there are three main designs of multicore fibers:

a. Simple homogeneous MCOFs,

b. Homogeneous MCOFs with a barrier layer,

c. Heterogeneous MCOFs.

In general, simple homogeneous MCOFs are multicore optical fibers with the same core parameters, that is, with the same core diameters and the same difference in the refractive indices between the core and cladding. The key parameter for each MCOF is the value of optical crosstalk, since when an optical beam propagates along it, there is a probability of interference of the mode fields of neighboring cores, leading to the transition of radiation from one core to another, or, in other words, to optical crosstalk [35, 36]. It is clear that the level of this crosstalk depends largely on the MCOF design and the realistic operating conditions.

Then, we consider homogeneous MCOFs with a barrier layer. The main goal of most works aimed at developing the MCOF design was to make it with as many cores as possible. This creates limitations consisting, on the one hand, of the interference effect between the modes of neighboring cores, which leads to crosstalk. It is obvious that the closer the cores are, the stronger the coupling between them. Therefore, they must be located at a sufficiently large distance. On the other hand, if they are sufficiently far from each other, in order to increase the number of cores, it is necessary to increase the outer diameter of the MCOF. In this case, the critical diameter at which mechanical strength is not lost, as noted above, is from 200 to 220 μm.

In connection with the above, the studies aimed at reducing crosstalk for given intercore distances and core parameters. Let us assume that there are two cores with centers located at some distance from each other with an unacceptably high level of crosstalk. One promising way to reduce the mutual influence between the modes of neighboring cores is to introduce a region with a reduced refractive index into their profile (see **Figure 4**) [37].

Due to a barrier layer, located at some distance from the core, the electric field strength at the edge of the mode field distribution decreases. Thus, the mode field overlap is also reduced, leading to a decrease in crosstalk. At the same time, a layer with a reduced refractive index, located at some distance from the core, does not significantly affect the mode properties of individual cores. Because of reducing the electric field strength at the edge of the mode field distribution, the overlap integral of the fields of the two modes decreases, and, consequently, the core coupling coefficient and crosstalk. An example of a design of an MCOF with seven identical cores arranged in a hexagonal pattern, each of which is surrounded by a barrier layer with a reduced refractive index, is shown in **Figure 5** [38]. The profile of the change in the refractive index in one of the cores and the barrier layer is shown in **Figure 6**.

For comparison, the authors of that publication also manufactured a MCOF without a barrier layer (MCOF-C). The parameters of all three MCOFs (A, B, and C) are given in **Table 6**. As can be seen, the presence of a barrier layer in this MCOF design allowed us to reduce optical crosstalk by approximately 20 dB over a length of 1 km (from minus 34 dB in MCOF-C to minus 55 dB in MCOF-A).

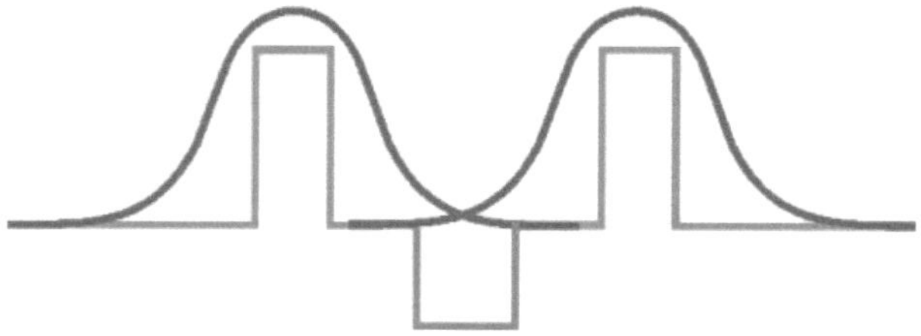

Figure 4.
The refractive index profile (gray line) and the electric field strength distribution (orange line) of two cores between which a barrier layer with a reduced refractive index is located.

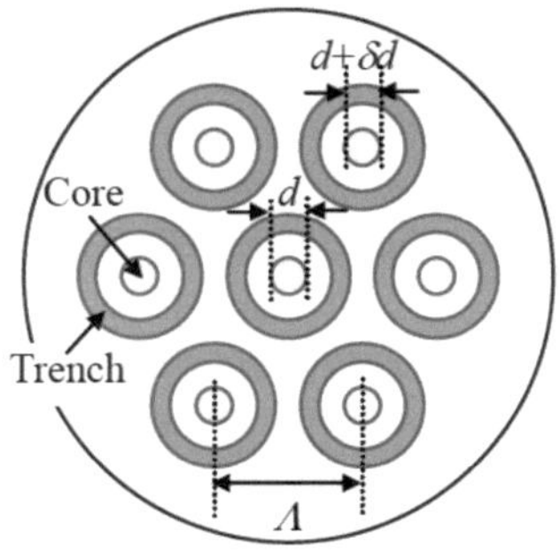

Figure 5.
Cross-section of homogeneous MCOF with barrier layer [38].

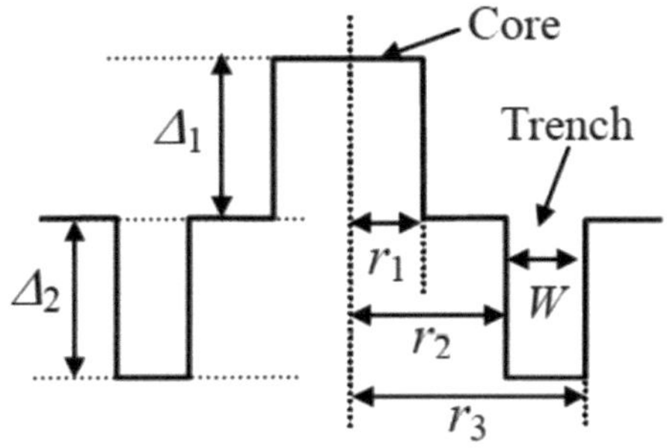

Figure 6.
The refractive index profile of one of the cores and the barrier layer.

Another way to reduce optical crosstalk is phase mismatching between the modes of adjacent cores, that is, using cores with different effective refractive indices (propagation constants) of the modes [39]. This can be done, for example, if two adjacent cores have the same diameter but different refractive indices. Such MCOFs are called heterogeneous [40]. **Figure 7** schematically shows the refractive index profile of two adjacent cores. The modes of different cores have different effective refractive indices n_{eff}, which are related to the mode propagation constant β, as $\beta = kn_{neff}$. That is, the mode propagation constants of the two cores β_1 and β_2 are also different, and the greater the difference between the effective refractive indices of adjacent modes, the less the crosstalk, and even a very small difference in them should lead to a significant reduction in crosstalk.

However, this type of MCOF design does not provide a complete solution to the problem of reducing optical crosstalk since the magnitude of the phase mismatch is affected by bends, which are always present in the laid MCOF and have an uncontrolled nature. As is known, a curved FOL is equivalent to a straight one with an inclined refractive index profile [41]. To illustrate this effect, **Figures 7** and **8** show the effective refractive index profiles for both versions of the heterogeneous MCOF. As follows from the Figures, even if the core diameters are the same, the effective refractive indices of the modes of these two cores, n_{eff1} and n_{eff2}, and, consequently, the propagation constants will be different since the effective refractive index profile changes as a result of bending. Moreover, as follows from **Figure 8**, at a certain bending radius, a complete coincidence of the effective refractive indices of the modes can occur.

Parameter	MCOF-A	MCOF-B	MCOF-C
The difference in refractive indices of the core and the cladding	0.0054	0.0052	0.0058
Difference in refractive indices of the cladding and barrier layer	−0.0106	−0.0102	—
r_1 (**Figure 6**), μm	4.1	4.1	4.04
r_2/r_1 (**Figure 6**)	2,23	2,23	—
r_3/r_1 (**Figure 6**)	3,23	3,23	—
W/r_1 (**Figure 6**)	1	1	—
Number of cores	7	7	7
Distance between the centers of adjacent cores, μm	38.3	35.4	39.4
MCOF diameter, μm	136.9	125.4	137.5
Mode field diameter (wavelength 1.31 μm), μm	8.7	9.2	8.6
Measured optical crosstalk value (wavelength 1.55 μm), dB/km	−55	−45	−34

Table 6.
The MCOF under test parameters [38].

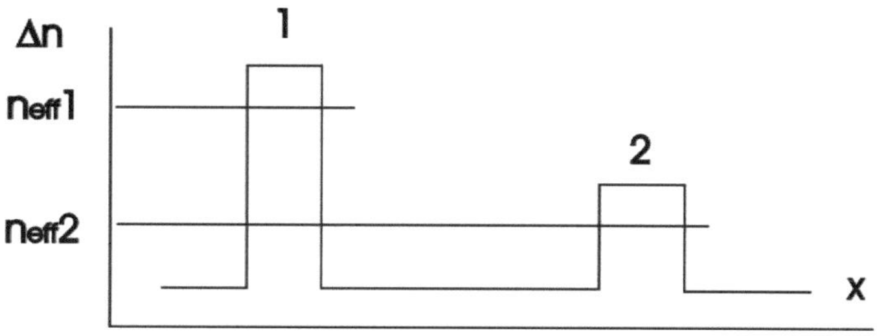

Figure 7.
Example of refractive index profile of a heterogeneous MCOF.

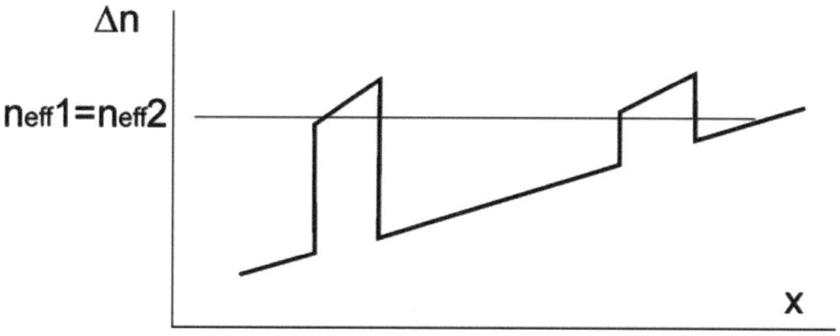

Figure 8.
Equivalent refractive index profile of curved heterogeneous MCOF.

Thus, if in the MCOF with different parameters of adjacent cores, in the straight state a small value of crosstalk is observed due to a significant difference in the effective refractive indices of the modes, in the bent position of the MCOF they can increase due to the convergence of the effective refractive indices of the modes (see **Figure 8**). The smaller the permissible bending radius, the greater the difference in the effective refractive indices, which is necessary to avoid the occurrence of the phase-matching condition. Therefore, the use of phase mismatch of the modes of adjacent cores can be ineffective in terms of reducing crosstalk, especially if the MCOF is bent with a sufficiently small radius. In addition, the change in the value of crosstalk associated with a change in the difference in the propagation constants of the modes due to bending is also difficult to control in an operating FOTS since it is impossible to clearly control the radii of the OF microbends in the optical cable [42]. The effect of a sharp increase in crosstalk during bends was observed in a number of experimental studies [43–45].

4.2 Approaches to reduce crosstalk in MCOF

A design option that avoids the increase in crosstalk when bending a heterogeneous MCOF is to use a rectangular cross-section of the MCOF with cores arranged in one row, as shown in **Figure 9** [46]. As follows from the Figure, due to the fact that the MCOF with a rectangular cross-section is bent mainly around an axis parallel to its long side, since such a bend does not reduce the difference in the effective refractive indices of the modes for adjacent cores, therefore, there will be no effect of increasing optical crosstalk.

In Ref.s [47, 48], the design of a standard circular cross-section of MCOF is proposed in which crosstalk is reduced by reducing the number of surrounding neighboring cores. This can be either a structure with a ring-shaped arrangement of cores (**Figure 10**) or a structure with a ring-shaped arrangement of side cores and a distance from the center of the side core to the center of the central core that is much

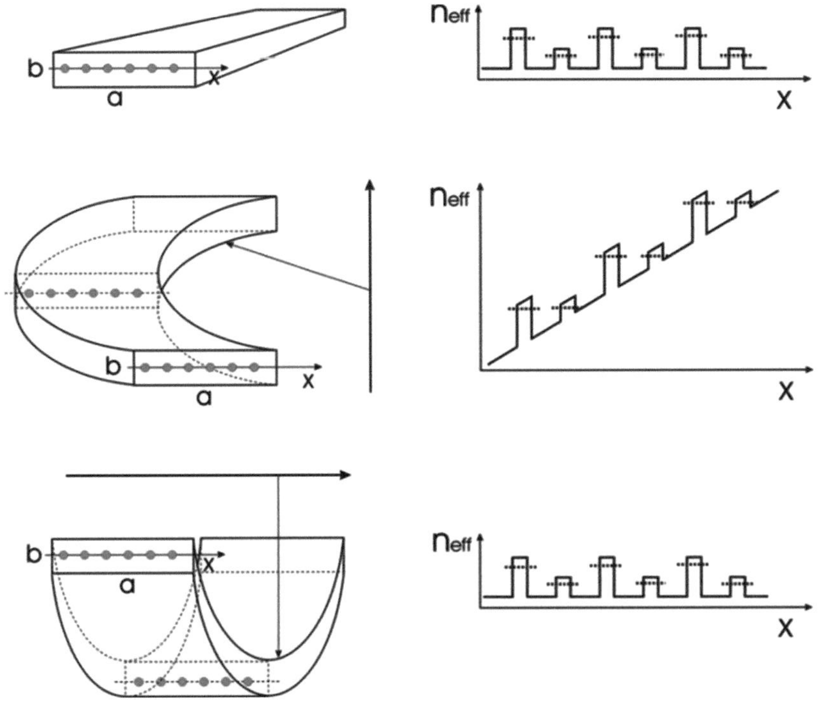

Figure 9.
A rectangular cross-section of MCOF bent in different directions (left) and the corresponding equivalent refractive index profile (right) [46].

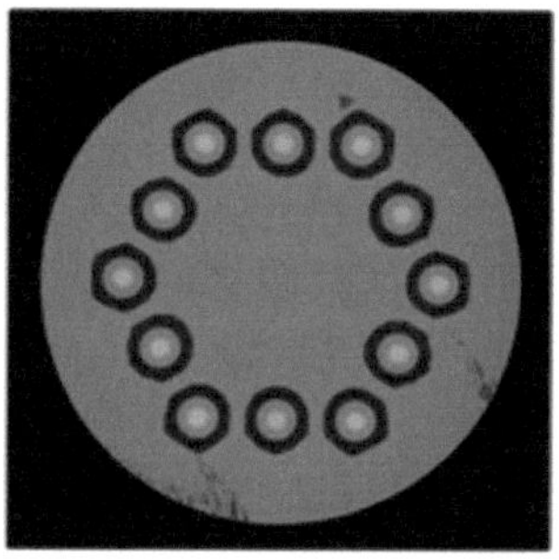

Figure 10.
Cross-section of MCOF with annular core arrangement [48].

greater than between the centers of the side cores (**Figure 11**). This arrangement of cores makes it possible to reduce crosstalk in the central core and increase the number of cores in an MCOF with a diameter of about 200 μm. For example, in an MCF with a diameter of 204 μm, ten cores can be placed [48] as opposed to seven cores in a hexagonal packing. In this case, the distance between the centers of the side and central cores was 59.2 μm, and between the centers of the side cores—40.5 μm with an outer diameter of 204.4 μm. In the studied design, with a mode field diameter of 12.4 μm at a wavelength of 1.55 μm, the optical crosstalk between the side cores was minus 46 dB/km, and between the central and side cores—minus 76 dB/km.

Also, in Ref. [47], the designs of the MCOF with a linear arrangement of cores are proposed (see **Figure 12**). The main potential advantage of such MCOF is its use in signal transmission paths in lower-scale data centers since the geometry of the

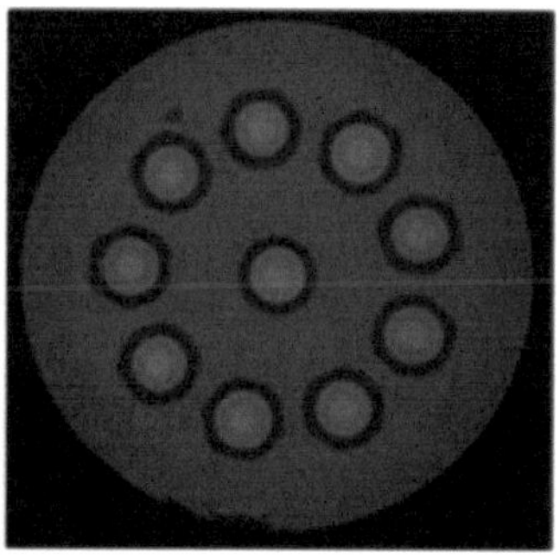

Figure 11.
Cross-section of MCOF with different distances between the side cores and between the side and central cores [48].

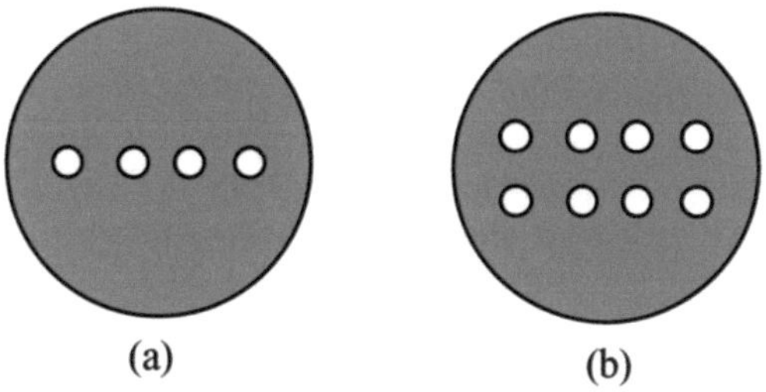

Figure 12.
Cross-sections of MCOF with a linear arrangement of cores [49].

Model	Manufacturer	Number of cores	Core arrangement	Cladding diameter (μm)	Site
SM-4C1500	Fibercore, GB	4	Square	125 ± 1	https://fibercore.humaneticsgroup.com/products/multicore-fiber
SM-7C1500	Fibercore, GB	7	Hexagonal with central core		
IXF-MC-12-PAS-6	iXblue, France	12	Dodecagonal	355 ± 15	https://www.ixblue.com/photonics-space/multicore-fibers/
IXF-MC-7-SM-1550	iXblue, France	7	Hexagonal with central core	245 ± 15	
IXF-C4-EDF-FGC-980	iXblue, France	4	Square	245 ± 15	
To order	OFS Furukawa, USA	From 4 to 8	—	—	https://www.ofsoptics.com/multicore-optical-fiber/
To order	Sumitomo Electric, Japan	From 8 to 12	—	—	https://sumitomoelectric.com/rd/optical-communications-laboratory/multi-core-fiber
To order	Chiral Photonics, USA	From 4 to 7	—	—	https://chiralphotonics.com/products/multicore-fiber-fanout/

Table 7.
Examples of modern global production of MCOF.

arrangement of the cores repeats the geometry of the arrangement of OTC devices, which simplifies their joining.

4.3 Examples of MCOF production

Currently, a number of well-known global manufacturers of fiber-optic components and systems have organized the pilot production of single-mode MCOFs. The results of our search for examples of MCOF operating in the C-band (from 1530 to 1565 nm) are presented in **Table 7**.

As follows from the Table, the MCOF technology has already been mastered by a number of well-known companies in Europe, Asia, and North America, but their widespread production is still absent. The main reason, most likely, is the current underdevelopment of FOTS with SpMP compared to TMP and SMP. Nevertheless, as discussed above, it is planned to change the situation as the global need for the development of FOTS with terabit and petabit throughput expands.

4.4 Simulation

When describing mathematically a transient noise of two MCOF optical waveguides, the Coupled Modes Theory is used. According to it, in axisymmetric optical

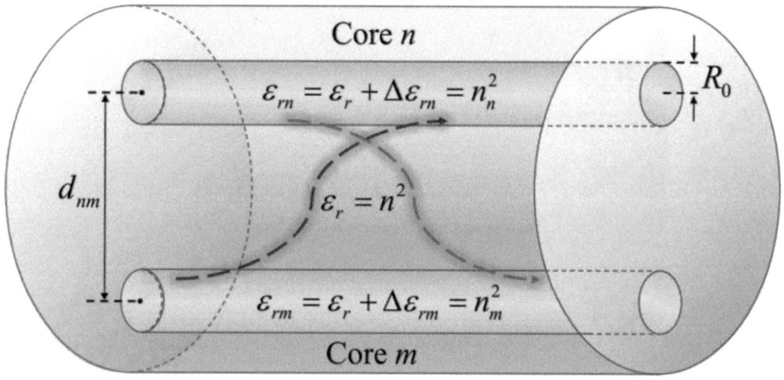

Figure 13.
Diagram of a double-core lightguide.

fibers there is a set of propagating modes. If two such fibers are located close enough, the modes in them will begin to interfere with each other. If the distributions of the electromagnetic field before and after the interaction of the modes do not differ significantly, then the waveguide properties of such fibers can be considered using perturbation theory [50]. In it, the interaction between cores n and m of a MCOF can be considered as field propagation in a double-core fiber made up of these cores (see **Figure 13**).

This model assumes the absence of nonlinear effects associated with silica glass. In addition, neglecting the effect of birefringence, we can assume that the polarization in both cores is the same, and the interaction occurs only between modes that are non-orthogonal in polarization. Then, the complex amplitude of the propagating electric field can be described as [50–53]:

$$\vec{E}_g\left(\vec{r}\right)\approx\vec{E}_n\left(\vec{r}\right)+\vec{E}_m\left(\vec{r}\right)=\begin{bmatrix} A_n(z)F_n(r,\varphi)\exp\left(-j\beta_n z\right) \\ +A_m(z)F_m(r,\varphi)\exp\left(-j\beta_m z\right)\end{bmatrix}\cdot\hat{\tau}, \quad (1)$$

where, $A_{n(m)}$ – complex envelope; $F_{n((m)}$ – distribution of the LP_{01} mode field in the core and cladding; $\beta_{n(m)}$ is the propagation constant of the LP_{01} mode for the case of an unperturbed fiber; $\hat{\tau}$ – unit polarization vector. Let us substitute into Maxwell's equations:

$$\Delta\vec{E}_g\left(\vec{r}\right)+k_0^2\left[\varepsilon_{\mathrm{rg}}\left(\vec{r}\right)\right]\vec{E}_g\left(\vec{r}\right)=\vec{0}, \quad (2)$$

where, ε_{rg} is the relative electrical permittivity.

$$\varepsilon_{\mathrm{rg}}\left(\vec{r}\right):=\varepsilon_r\left(\vec{r}\right)+\Delta\,\varepsilon_{\mathrm{rn}}\left(\vec{r}\right)+\Delta\,\varepsilon_{\mathrm{rm}}\left(\vec{r}\right). \quad (3)$$

In addition, for each core the Helmholtz equation must be satisfied:

$$\Delta_T F_{n\,(m)}+k_0^2\varepsilon_{\mathrm{rg}}F_{n\,(m)}=\left\lfloor k_0^2\Delta\varepsilon_{rm\,(n)}+\beta_{n\,(m)}^2\right\rfloor F_{n(m)}. \quad (4)$$

Combining all expressions, we get:

$$\begin{bmatrix} j\left[\dfrac{dA_n}{dz}+X_{nm}\exp(-j\Delta\beta_{\mathrm{mn}}z)\right]\dfrac{dA_m}{dz} \\ =c_nA_n+k_{\mathrm{nm}}\exp(-j\Delta\beta_{\mathrm{mn}}z)A_m \end{bmatrix}, \tag{5}$$

where, $\Delta\beta_{mn} = \beta_m-\beta_n$; k_{nm} is the coupling coefficient between cores n and m.

$$k_{\mathrm{nm}}=\frac{k_0^2}{2\beta_n}\frac{\iint \Delta\varepsilon_{\mathrm{rn}}F_mF_n dS_\infty}{\iint F_n^2 dS_\infty}=\frac{k_0^2NA_n^2}{2\beta_n}\frac{\iint F_mF_n dS_n}{\iint F_n^2 dS_\infty}. \tag{6}$$

Next, to calculate transient losses (crosstalk), we will use the Coupled Power Theory. This approach is based on the principle of measuring signal power, where the power introduced into one core at the input of a fiber is transferred to and out of neighboring cores. The equation of related powers can be written as [54]:

$$\frac{dP_P}{dz}=\sum_{p\neq q}h_{pq}(z)\left[P_q(z)-P_p(z)\right],, \tag{7}$$

where, P_p, Pq – average signal powers in cores p, q; h_{pq} – power coupling coefficient between cores. In the case of identical cores ($\Delta\beta = 0$), solving the coupling equations gives an expression for the power-coupling coefficient [54]:

$$\boldsymbol{h_{nm}=\frac{2k_{nm}^2R_b}{\beta_m\Lambda}} \tag{8}$$

Taking this into account, crosstalk from the core n to the core m results in:

$$\boldsymbol{XT_{nm}=\tanh\left(h_{nm}L\right)=\tanh\left(\frac{2k_{nm}^2R_bL}{\beta_m\Lambda}\right)}, \tag{9}$$

where, R_b is the bending radius of the fiber; L – length of the light guide; Λ – distance between the centers of the cores. The above expression is true for the case of interaction of two cores. If any core has n equivalent neighbors, then the crosstalk induced to this core can be estimated as [55]:

$$\boldsymbol{XT\_worst\,[dB]=XT\_nm\,[dB]+10lg(n)}, \tag{10}$$

Finally, a core design with a trench of low-index glass can be used to suppress crosstalk (See **Figure** 6). The crosstalk in such a core can be estimated by knowing the crosstalk in the core without a trench [54]:

$$XT_{trench}[dB]=XT_{step}[dB]+10\lg\Gamma-17.4(W_2-W_1)\frac{W}{r_1}, \tag{11}$$

where, $\boldsymbol{\Gamma=W_1/\left[W_1+\frac{(W_2-W_1)W}{\Lambda}\right]}$; $\boldsymbol{W_1\approx 1.1428V_1-0.996}$ for $1.5\le V_1\le 2.5$; $\boldsymbol{W_2=W\sqrt{V_2^2+W_1^2}; V_{1(2)}=2\pi r_1 n_{clad}\frac{\sqrt{2\Delta_{1(2)}}}{\Lambda}}$.

Thus, knowing the distribution of the mode field in each of the cores separately, it is possible to calculate the crosstalk introduced by them in any MCOF.

As part of crosstalk simulation, a model of a MCF consisting of 19 identical single-mode cores was considered. The necessary initial data for the calculations are given in **Table 8**.

The calculations compared the three most commonly used prospective 19-core MCOF designs: ring, double hexagonal, and double ring. Their location diagrams are shown in **Figure 14**.

To determine the introduced crosstalk, first of all, it is necessary to calculate the mode field distribution of a pair of cores, for which the Comsol Multiphysics software package was used. With its help, the fundamental mode field of each core was first calculated separately, and then the overlap integral between the fields was calculated.

The calculation showed the following results. In the case of a ring core arrangement (**Figure 14a**), the largest crosstalk of −13.7 dB is induced between adjacent cores in the outer ring. The disadvantage of this scheme is the small residual thickness of the cladding. So, to level out the effect of fiber micro-bending on losses, it should be at least 30 μm. Also, in the case of a double hexagonal core arrangement (**Figure 14b**), the largest crosstalk of −8.9 dB is induced. Unlike the previous scheme, here a large number of cores simultaneously experience significant interference from neighboring ones. For example, six cores affect the central core at once, causing significant interference. However, in this design the residual cladding thickness is greater, which allows the cores to be moved further apart. Finally, in the case of a double-ring core

Wavelength, λ	**1.55 μm**
Cladding refractive index	1.45
Core refractive index	1.4551 ($\Delta = 0.35\%$)
Core diameter	9 μm
Cladding diameter	250 μm
Minimum distance between cores, Λ	40 μm
Maximal bend radius, R_b	140 mm
Fiber length, L	5 km

Table 8.
Main parameters of the MCOF under test.

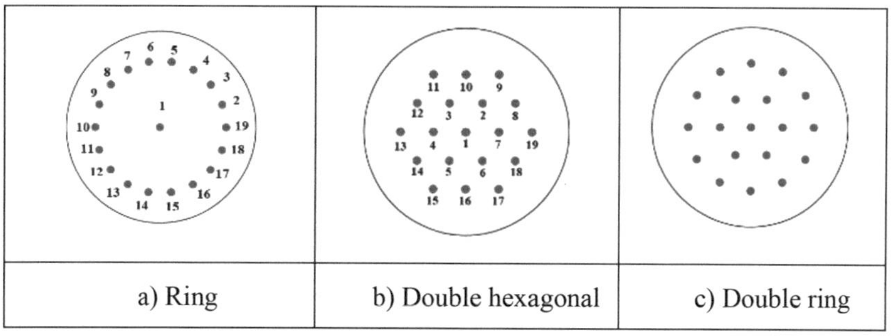

Figure 14.
Core location diagram in 19-core MCOF under test. a) Ring, b) Double hexagonal, c) Double ring.

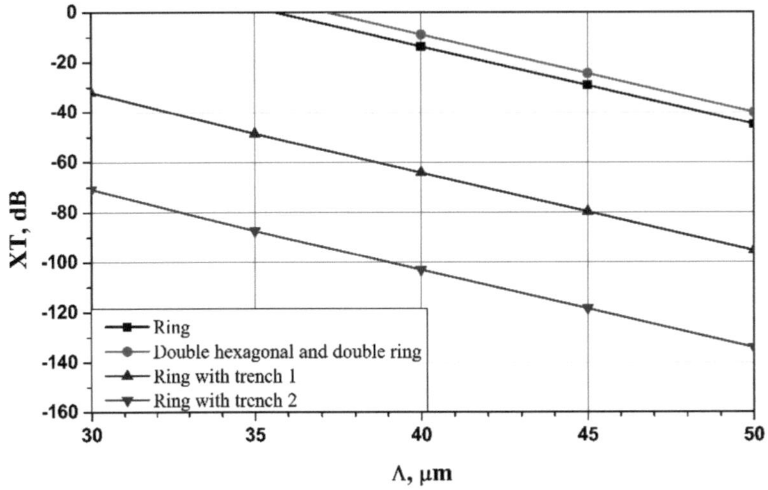

Figure 15.
Results of calculating the crosstalk levels.

arrangement (**Figure 14c**), the largest crosstalk of −8.9 dB is induced too. In this design, the greatest crosstalk is induced at the inner ring cores. Compared to the single ring design of **Figure 14a**, this one allows for a greater thickness of the outer cladding to be left. Among the disadvantages of such a scheme, it should be noted that there is greater crosstalk on the inner ring of cores than in a double hexagonal scheme.

The results of the study in the form of crosstalk (XT) as a function of the minimum distance between the cores (Λ) for their various configurations (**Figure 14**) and internal core design (**Figure 6**) are presented in **Figure 15**.

As **Figure 15** shows, the use of homogeneous cores results in unacceptably high levels of crosstalk in all three core configurations. In particular, with a typical core spacing of 40 μm, it exceeds −20 dB, which will lead to a significant deterioration in the quality of transmitted optical signals. An effective way to overcome this situation is to use lower refractive index trenches around the core (see **Figures 5** and **6**). Namely, depending on the degree of their reduction, the crosstalk level can be reduced to −60 dB or even −100 dB, which will ensure transmission quality compared to the single-core fiber level.

4.5 Prototyping and experiments

From **Table 7**, it can be concluded that the main efforts of the world's optical fiber manufacturers are aimed at creating 7-core MCOFs, which is associated with the above-described development of FOTS for 5G cellular telecommunication systems (see Subsection 3.1), as well as for optical connecting links for hyper-scale data centers (see Subsection 3.2). However, there is still no clear solution for such an important parameter from the point of view of the permissible level of crosstalk interference as the optimal cladding diameter. In particular, in **Table 7**, it changes from the standard equal to 125–245 μm for a 7-core MCOF or to 355 μm for a 12-core MCOF. Another feature of the MCOFs listed in **Table 7** is the use of a fundamentally new manufacturing technology, which has led to their significantly higher cost compared to single-core analogs. However, for MCOFs with a relatively small number of cores, it is possible to use a much simpler technology based on the processing and sintering of single-core preforms, the number of which coincides with the required number of cores [55]. Following this, the results of a comparative experimental study of crosstalk for two 7-core simple homogeneous MCOF prototypes, each about 1 km long, are presented below: with a standard cladding diameter of 125 μm (Prototype 1) and with one increased to 200 μm (Prototype 2).

The testing of both designed and homemade MCOF prototypes was carried out according to the following approach. For measurement possibility, both MCOF prototypes were terminated with homemade 7-core input/output devices based on seven short sections of a single-mode single-core OF with an optical connector of the FC/APC type at one of the ends. The other end was adjusted for minimum input losses, after which it was rigidly connected to the output of the corresponding MCOF core using glue and optical welding. During the measurement of the core, the following designation of the cores was used (see **Figure 16**).

Comparative testing of both prototypes consisted of measuring two critical parameters in the third transparency window of silica OF (in the region of 1.55 μm): transmission losses and crosstalk.

First, optical losses were tested. The measurements were carried out according to the standard method using an optical reflectometer. The measurement results with the numbering of cores according to **Figure 16** are presented in **Table 9**.

Then, crosstalk testing was performed on the both 7-core prototypes. The testing was carried out using the standard method of measuring the difference in optical power by injecting a 20-mW laser emission into the corresponding core and determining the output power level of the corresponding core using an optical power meter. The measurement results are shown in **Table 10**.

The results of the comparative testing (**Tables 9** and **10**) allows us to draw the following outcomes:

1. In the Prototype 1 with a standard cladding diameter of 125 μm, the obtained values of optical losses at a wavelength of 1.55 μm are from 3.5 to 5.8 dB/km in the side cores and slightly above 2 dB/km for the center core, that is, they have increased significantly compared to the permissible values of optical losses of a single-core optical fiber, which, according to the G.652A standard, should not exceed 0.4 dB/km. On the contrary, in Prototype 2, with a cladding diameter

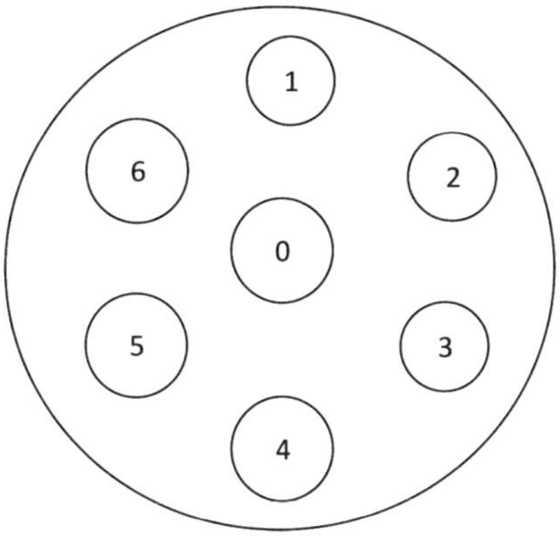

Figure 16.
Prototype core designations.

Core number	Prototype 1, dB/km	Prototype 2, dB/km
0	2.2	2,0
1	3,48	2,1
2	5,71	2,8
3	4,08	3,3
4	5,78	3,4
5	5,44	3,1
6	5,68	3,0

Table 9.
Results of measuring optical losses in the cores of prototypes.

Core number	Prototype 1		Prototype 2	
	Crosstalk between adjacent side cores, dB	Crosstalk relative to the central core, dB	Crosstalk between adjacent side cores, dB	Crosstalk relative to the central core, dB
0-1	—	−28	—	−64.2
1-2	−15	−21	−52.9	−58.9
2-3	−14	−25	−55.7	−63.1
3-4	−17	−24	−51.6	−62.3
4-5	−17	−24	−52.6	−60.5
5-6	−14	−22	−53.3	−63.3
6-0	—	−26	—	−64.6

Table 10.
Results of measuring optical crosstalk in the cores of prototypes.

increased to 200 μm, the obtained values of optical losses at a wavelength of 1.55 μm at the level of 2 dB are in the central core and from 2.1 to 3.4 dB in the side cores, that is, significantly less. Moreover, the increased losses compared to the standard SCOF are most likely explained by additional losses introduced by the homemade input/output devices installed at both ends of the MCOF prototype.

2. The results of testing Prototype 1 showed a significant deterioration in signal transmission quality due to unacceptably high crosstalk, ranging from −14 to −17 dB for adjacent side cores and ranging from −21 to −28 dB relative to the central one. On the contrary, in Prototype 2, the same parameter for both options is in the range from −51 to −64 dB, providing signal transmission quality comparable to single-core OF.

3. The conducted analysis of the common causes of such different values of critical parameters in both studied prototypes of the MCOF under testing allowed us to draw the following conclusions. So, our studies have revealed the following reasons for the effect of increased losses.

a. The increased value of losses in the central core is obviously associated with the too small value of the so-called normalized operating frequency or V-parameter of the light guide, which is determined from the expression:

$$V = \frac{2\pi a}{\lambda} NA, \tag{12}$$

where α is the core radius; λ is the operating wavelength; *NA* is the aperture of the light guide. A consequence of this is high sensitivity, even to micro-bending.

a. In the side cores, the value of optical losses is higher than in the central core, which is associated with the too close location of the cores to the outer coating of the lightguide in combination with the small value of the V-parameter. Most likely, the increase in losses occurs because the edge of the transverse distribution of the mode radiation penetrates the polymer coating with a refractive index greater than that of silica glass, resulting in mode leakage into the cladding.

b. The crosstalk value was in the range of −14 ... −20 dB, which is significantly greater than can usually be obtained for a given distance between the cores with core parameters corresponding to the G652 standard. This is due to the large spreading of the core mode fields due to the small value of the V-parameter. The insufficient value of the V-parameter is indicated by the underestimated cutoff wavelength of the first higher mode.

5. Conclusion

In the chapter, following modern trends of reaching a petabyte capacity in the development of fiber-optic telecommunication systems, including 5G NR networks for megalopolises and interconnection links of hyper-scale data centers, an original principle and a conceptual block diagram of its design are proposed, including terminal transceivers using worldwide known time and spectral multiplexing and demultiplexing and an optical path using a relatively new low energy consumption spatial multiplexing and demultiplexing based on a multicore optical fiber. The results of the simulation, prototyping, and comparative experimental studies showed that the proposed way makes it possible to the creation of a telecommunications fiber-optic system for the abovementioned applications with high technical and economic efficiency, with a capacity of up to 10 Tbps even when using time multiplexing at the generally applicable level of 40 Gbps, up to 8-channel spectral multiplexing, and 7-core spatial multiplexing. Our further work in this topic will be directed on developing a 19-core optical fiber path, which, along with the use of a larger number of spectral channels, will make it possible to create low-power fiber-optic telecom systems with a throughput of up to hundreds of Tbps.

Acknowledgements

This work was funded by the Russian Ministry of Higher Education and Science in the framework of grant FSFZ-2022-0005.

Conflict of interest

The authors declare no conflict of interest.

Author details

Mikhail E. Belkin[1*], Sergei L. Semjonov[2], Olga N. Egorova[2], Eugeniy Plastinin[2], Mikhail G. Vasil'ev[3] and Alexander S. Sigov[1]

1 MIREA - Russian Technological University, Moscow, Russian Federation

2 Prokhorov General Physics Institute of the Russian Academy of Sciences, Moscow, Russian Federation

3 Kurnakov Institute of General and Inorganic Chemistry, Russian Academy of Sciences, Moscow, Russian Federation

*Address all correspondence to: belkin@mirea.ru

References

[1] 5G mobile communications systems for 2020 and beyond. 5GMF white paper. In: Fifth Generation Mobile Communications Promotion Forum (5GMF). 2017. 250 p. Available from: https://5gmf.jp/wp/wp-content/uploads/2016/08/5GMF_WP100_Summary-E.pdf

[2] Lisa H. Data center interconnect strategies. In: Lightwave. Editorial Guide; 2020. 29 p. Available from: https://www.blog.adtran.com/en/data-center-interconnect-strategies-part-1

[3] Bell Telephone Laboratories. Transmission Systems for Communications. 4th ed. North Carolina: Incorporated Winston-Salem; 1971. 769 p

[4] Lee BG, Kang M, Lee J. Broadband Telecommunications Technology. 2nd ed. Boston, London: Artech House Inc.; 1996. 658 p

[5] Corning. Available from: https://www.corning.com/ru/ru.html

[6] Freeman RL. Fiber-Optic Systems for Telecommunications. 3rd ed. John Wiley & Sons, Inc.; 2002. 440 p

[7] Gowar J. Optical Communication Systems. London: Prentice-Hall International, Inc.; 1984. 483 p

[8] Agrawal GP. Fiber-Optic Communication Systems. Wiley; 1997. 545 p

[9] Bailey D, Wright E. Practical Fiber Optics. USA: Elsevier; 2003. 288 p

[10] Paschotta R. Field Guide to Optical Fiber Technology. SPIE Press; 2010. 117 p

[11] Optical Line Systems GX Series Compact Modular Platform. Available from: https://www.infinera.com/products/gx-series-compact-modular-platform/

[12] Schmidt T, Malouin C, Liu S. Recent trends in 100G module and subsystem development for long haul DWDM applications. In: Optoelectronics and Communications Conference, 2009. Hong Kong, China. 2009. pp. 1-2

[13] Matsuo S, Takenaga K, Saitoh K, Nakajima K, Miyamoto Y, Morioka T. High-spatial-multiplicity multi-core fibres for future dense space-division-multiplexing system. In: European Conference on Optical Communication ECOC'2015. 2015. pp. 1464-1475

[14] Andrews JG, Buzzi S, Choi WS, Hanly SV. What will 5G be? IEEE Journal on Selected Areas in Communications. 2014;**32**:1065-1082

[15] Chen S, Zhao J. The requirements, challenges and technologies for 5G of terrestrial mobile telecommunication. IEEE Communications Magazine. 2014; **52**:36-43

[16] Belkin ME, Bakhvalova T, Sigov AC. Design principles of 5G NR RoF-based fiber-wireless access network. In: Recent Trends in Communication Networks. Vol. 1. London, UK, London, UK: IntechOpen; 2020. pp. 121-141. Available from: https://www.intechopen.com/chapters/70248

[17] Belkin ME, Zhukov L, Sigov AC. Predesigning RoF-based millimeter-wave access sub-network in megapolis 5G communication system. In: Proceedings of International Conference on Microwaves, Communications, Antennas and Electronic Systems, COMCAS 2024, Tel Aviv, Israel, 9-11 July 2024. 2024. 6 p

[18] Cisco Global Cloud Index: Forecast and Methodology, 2016-2021. Cisco White Paper; 2018

[19] Savvas A. Optics in the Data Centre. Optical Connections, Q1, No. 32. 2023. pp. 10-12

[20] The Role of Data Centers in an Interconnected World. Available from: https://withoutyou.de-cix.net/the-role-of data-centers/

[21] Thiry M. ABB Decathlon® for Data Centers Partner Strategy [Electronic Resource]. 2021. Available from: https://slideplayer.com/slide/14007994/

[22] Rack Density Keeps Rising at Enterprise Data Centers [Electronic Resource]. 2021. Available from: https://datacenterfrontier.com/rack-density-keeps-rising-at-enterprise-data-center

[23] Rofoee B et al. Programmable on-chip and off-chip network architecture on demand for flexible optical intra-datacentres. Optics Express. 2013;**21**: 5475-5480

[24] Michel S. Data Center Ethernet on the Move to 224 Gbps [Electronic resource]. Available from: https://blogs.keysight.com/blogs/inds.entry.html/2021/12/01/data_center_ethernettechnologyonthemoveto224-TfjK.html

[25] Yan Y et al. All-optical programmable disaggregated data centre network realized by FPGA-based switch and interface card. IEEE OSA Journal of Lightwave Technology. 2016;**34**:1925-1932

[26] Belkin ME, Plastinin E, Smirnov N. Designing a low-energy fiber-optic interconnect link of terabyte throughput for hyper-scale data centers. In: Proceeding of International Multidisciplinary Modeling & Simulation Multiconference, I3M 2024, Tenerife, Spain, 18-20 September 2024. 2024. 8 p

[27] Recommendation ITU-T G.692

[28] Zhu B, Taunay TF, Yan MF, Fini JM, Fishteyn M, Monberg EM, et al. Seven-core multicore fiber transmissions for passive optical network. Optics Express. 2010;**11**:11117-11122

[29] Rosinski B, Chi JWD, Grosso P, Le Bihan J. Multichannel transmission of a multicore fiber coupled with vertical-cavity surface-emitting lasers. Journal of Lightwave Technology. 1999:807-810

[30] Inao S, Sato T, Sentsui S, Kuroha T, Nishimura Y. Multicore optical fiber. In: Proc. Optical Fiber Communication Conference. 1979. p. WB1

[31] Qian D et al. 101.7-Tb/s (370x294-Gb/s) PDM-128QAM-OFDM transmission over 3 × 55-km SSMF using pilot-based phase noise. In: Proceedings of Optical Fiber Communication Conference. 1979. p. PDPB5

[32] Takara H et al. 1.01-Pb/s (12 SDM/222 WDM/456 Gb/s) crosstalk-managed transmission with 91.4-b/s/Hz aggregate spectral efficiency. In: Proc. European Conference of Optical Communications. 2012. p. Th.3.C.1

[33] Puttnam BJ, Luis RS, Klaus W, Sakaguchi J, Mendinueta J-MD, Awaji Y, et al. 2.15 Pb/s transmission using a 22 core homogeneous single-mode multi-core fiber and wideband optical comb. In: Proceedings of the European Conference of Optical Communications. 2015. p. PDP 3.1

[34] Berthold J. Optical networking for data centers across wide area networks.

In: Proceedings of Optical Fiber Communication Conference. 2012. p. OW1J.1

[35] Dianov EM, Semjonov SL, Bufetov IA. New generation of optical fibers. Quantum Electronics. 2016;**46**(1):1-10. DOI: 10.1070. QE2016v046n01ABEH015963

[36] Egorova ON, Belkin ME, Zhuravlev SG, Semjonov SL. Bending effects in multicore optical fibers for fiber-optic delay line. In: Proceeding of PIERS Symposium, 2018, Toyama, Japan. 2018. 4 p

[37] Dianov EM, Semenov SL, Egorova ON. Multicore Optical Fiber. USA patent 9052433; 2015. claimed 19.01.2011; published 09.06.2015

[38] Takenaga K, Arakawa Y, Tanigawa S, Guan N, Matsuo S, Saitoh K, et al. Reduction of crosstalk by trench-assisted multi-core fiber. In: Proceedings of Optical Fiber Communication Conference. 2011. p. OWJ4

[39] Kumar S, Manyam UH, Srikant V. Optical Fibers Having Cores with Different Propagation Constants, and Method of Manufacturing the Same. USA patent 6611648; 2003. claimed 09.05.2001; published 26.08.2003

[40] Koshiba M, Saitoh K, Kokubun Y. Heterogeneous multi-core fibers: Proposal and design principle. IEICE Electronics Express. 2009;**6**(2): 98-103

[41] Marcuse D. Influence of curvature on the losses of doubly clad fibers. Applied Optics. 1982;**21**(23)

[42] Fini JM, Zhu B, Taunay TF, Yan MF. Statistics of crosstalk in bent multicore fibers. Optics Express. 2010;**18**(14): 15122-15129

[43] Hayashi T, Nagashima T, Shimakawa O, Sasaki T, Sasaoka E. Crosstalk variation of multi-core fibre due to fibre bend. In: Proceedings of European Conference of Optical Communications. 2010. p. We.8.F.6

[44] Hayashi T, Sasaki T, Sasaoka E, Saitoh K, Koshiba M. Physical interpretation of intercore crosstalk in multicore fiber: Effects of macrobend, structure fluctuation, and microbend. Optics Express. 2013;**21**(5):5401-5412

[45] Sasaki Y, Amma Y, Takenaga K, Matsuo S, Saitoh K, Koshiba M. Investigation of crosstalk dependencies on bending radius of heterogeneous multicore fiber. In: Proceedings of Optical Fiber Communication Conference. 2013. p. OTh3K.3

[46] Egorova ON, Semjonov SL, Senatorov AK, Salganskii MY, Koklyushkin AV, Nazarov VN, et al. Multicore fiber with rectangular cross-section. Optics Letters. 2014;**39**(7): 2168-2170

[47] Matsuo S, Sasaki Y, Ishida I, Takenaga K, Saitoh K, Koshiba M. Recent progress on multi-Core fiber and few-mode fiber. In: Proceedings of Optical Fiber Communication Conference. 2013. p. OM3I.3

[48] Matsuo S, Takenaga K, Arakawa Y, Sasaki Y, Taniagwa S, Saitoh K, et al. Large-effective-area ten-core fiber with cladding diameter of about 200 μm. Optics Letters. 2011;**36**(23):4626-4628

[49] Li M-J, Hoover B, Nazarov VN, Butler DL. Multicore fiber for optical interconnect applications. In: Proceeding of Optoelectronics and Communications Conference. 2012. p. 5E4-2

[50] Okamoto K. Fundamentals of Optical Waveguides. Elsevier; 2021

[51] Macho A., Morant M., Llorente R. Unified model of linear and nonlinear crosstalk in multi-core fiber. Journal of Lightwave Technology. 2016. 34. 13. C. 3035-3046.

[52] Koshiba M et al. Multi-core fiber design and analysis: Coupled-mode theory and coupled-power theory. Optics Express. 2011;**19**(26):B102-B111

[53] Saitoh K, Matsuo S. Multicore fiber technology. Journal of Lightwave Technology. 2016;**34**(1):55-66

[54] Ye F et al. Simple analytical expression for crosstalk estimation in homogeneous trench-assisted multi-core fibers. Optics Express. 2014;**22**(19): 23007-23018

[55] Egorova ON, Belkin ME, Klushnik DA, Zhuravlev SG, Astapovich MS, Semojnov SL. Microwave signal delay line based on multicore optical fiber. Physics of Wave Phenomena. 2017;**25**:289-292. DOI: 10.3103/S1541308X17040082

Chapter 6

A Brief Review of Recent Advances in the Use of Optical Fibres to Enhance Readout and Increase the Number of Physical Qubits in Superconducting Quantum Computers

Ricardo Rovere de Santi

Abstract

The promise of quantum computing has tantalised researchers for decades, offering the potential for a paradigm shift in computational power and cryptography security. However, progress has been painstakingly slow, hindered by the immense challenges implemented in such systems. Until recently, advancements in quantum computing were stymied by the necessity of operating at near-absolute zero temperatures, a requirement that severely constrained practical applications. Transmitting signals from room temperature to the cryogenic environment posed a significant bottleneck, limiting quantum processors to a mere thousand quits. But in 2021, a groundbreaking study titled "Control and readout of a superconducting quit using a photon link" introduced a transformation innovation—the replacement of traditional coaxial lines with optical fibres and modulated light. The use of optical fibres revolutionises quantum communication, providing a pathway to realising a million-quit computer. This chapter delves into the significance of these breakthroughs and elucidates the novel advancements achieved over the past 3 years, propelled by the original work and more, and will present a very brief introduction to quantum computer.

Keywords: quantum computing, optical fibre, cryogenics, superconductivity, quantum information

1. Introduction

The concept of the quantum Turing machine, initially proposed by Paul Benioff [1], evolved into a more advanced algorithm and machine idea by Richard Feynman and Yuri Mannin [2, 3], suggesting quantum computers could outperform classical computers. The feature which made the quantum computer theory more powerful is qubits in which electrons are described by quantum mechanics that have 0 (no charge) and 1 (charge) in superposition. The superposition can be written by $\psi = a_0|0\rangle + b_0|1\rangle$ where 0 and 1 are quantum states defined by vector where

$$|0\rangle = \begin{pmatrix} 1 \\ 0 \end{pmatrix} |1\rangle = \begin{pmatrix} 0 \\ 1 \end{pmatrix} \tag{1}$$

and a_0, b_0 are complex probability amplitude. In this form, the number of information compared to normal computer is $2^{qubit} = number\ of\ bits$. Therefore, using the principles of quantum mechanics, it is theoretically possible to increase the computation capability in orders of magnitude. Therefore, using the principles of quantum mechanics, it is theoretically possible to exponentially increase computational capabilities. However, many issues at the physical and engineering levels still need to be resolved. The complex nature of these challenges has hindered the development of such machines. Over the past two decades, amid stagnation in conventional electronic circuits, investment in quantum computing has surged. Since the mid-1980s, numerous algorithms have been developed, primarily leveraging black-box quantum computers. In the past decade, significant progress has been made in assembling quantum circuits using ion traps and superconducting materials, helping to elucidate the inner workings of possible quantum computers that surpasses the black box aspect of the theory. This has led to the formulation and implementation of more precise algorithms, yielding impressive results in areas such as exploring new material phases [4–6], molecular simulations [7], and simulating extreme stellar conditions. Furthermore, if quantum computers can be commercially developed, the quantum properties of qubits could revolutionise cryptography and information security [8]. Various devices and hardware proposals emerged in the first decade of the twenty-first century. During this period, a new set of guidelines, known as the DiVincenzo quantum computer criteria [9], was introduced to define the parameters for building physical quantum computers. These criteria have guided the development of numerous devices, including superconducting circuits [10, 11], trapped ions [12, 13], Rydberg atoms [14, 15], dipole molecules [16], semiconductors [17, 18], nuclear spins [19, 20], and photons [21, 22]. Despite these advancements, significant challenges remain in the creation of fault-tolerant, scalable quantum computers. This chapter focuses on the specific challenges related to the maximum number of qubit within superconducting circuits, for which partial solutions involving fibre optics have been explored.

1.1 Introduction to superconducting quantum computer

Superconducting qubits are solid-state electrical circuits that operate at temperatures below 20 mK, requiring extreme refrigeration to maintain these conditions. Typically housed in a room with a cryogenic refrigerator, these circuits rely on materials cooled to near absolute zero. At such low temperatures, materials exhibit key quantum properties, such as the elimination of electrical resistance and the significant reduction of thermal noise—both of which are crucial for the accurate measurement and manipulation of quantum amplitudes. In this environment, electrons behave according to the principles of quantum mechanics, where they can be described as wave-like entities characterised by complex probability amplitudes. These amplitudes define the likelihood of finding an electron in a particular quantum state, a foundational aspect of quantum computing. By harnessing these quantum behaviours, superconducting qubits allow for the manipulation of quantum states, enabling processes like superposition and entanglement. In conventional circuits at normal temperatures, the superposition or collapse of quantum waves averages out, making it impossible to observe quantum effects. However, in superconducting condensates, only waves with specific quantum energies are selected, enabling the direct observation of quantum phenomena. This transforms classical equations of energy in conductors and inductors into their quantum counterparts. Additionally, this approach extends Kirchhoff's laws into the quantum domain, resulting in a new set of equations that govern quantum circuits, analogous to their classical versions. The connection between classical and quantum circuits is established through the modulus square of the quantum amplitude, which corresponds to the classical charge distribution in the circuitry. By bridging these two realms, quantum circuits allow for a deeper understanding of electrical behaviour at the quantum level, with implications for both theoretical physics and practical quantum computing applications. Most quantum circuits are fabricated using advanced lithography techniques, which allow for precise patterning on a microscopic scale. These circuits are typically designed to operate in the radio-frequency (RF) spectrum, enabling the manipulation and measurement of quantum states with high precision. Materials such as niobium and tantalum [23, 24] are commonly used in their construction due to their excellent superconducting properties at cryogenic temperatures. Niobium, in particular, is favoured for its low critical temperature and high coherence, making it ideal for maintaining stable quantum states over extended periods. Tantalum has also gained attention for its robustness and ability to minimise decoherence, further enhancing the performance of quantum circuits. These materials, combined with lithographic precision, enable the creation of qubits that can reliably function within the demanding parameters of quantum computation. The design of these quantum circuits for the RF spectrum is crucial, as it allows for the efficient control and readout of qubit states. These quantum computing models have emerged as the leading technology in this field, with numerous governments and companies such as IBM, Microsoft, and Google investing billions of dollars into their development. Their success lies in their ability to create large and complex networks of qubits, offering four main advantages over other methods like trapped ions and Rydberg atoms. These four advantages are as follows.

One of the key advantages of superconducting qubits is the wide variety of qubit types that can be created. Different qubit designs, such as charge qubits, flux qubits, and phase qubits, can be tailored to specific applications. Critical parameters like qubit energy levels and coupling strength can be finely adjusted by modifying factors such as Josephson energy, inductance, and capacitance. These variables collectively define the total energy of the system, allowing for the construction of a unique Hamiltonian for each quantum circuit. The depiction of simplest quantum circuity and the associated energy and electronic components are displayed in **Figure 1**. Another major advantage of superconducting quantum computers is their ability to utilise the existing semiconductor infrastructure. The semiconductor industry has a highly advanced fabrication ecosystem, which facilitates the production of superconducting qubit circuits using components like capacitors, transistors, and inductors that are already well-established in conventional electronics. This synergy allows quantum computing to leverage over 50 years of technological advancements, enabling the production of smaller, more reliable, and higher quality quantum circuits. The ability to build on mature semiconductor manufacturing processes provides a significant edge in scaling up quantum systems and accelerating their development for

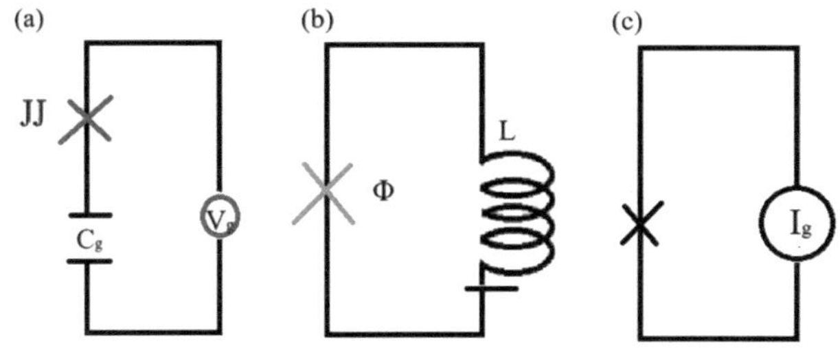

Figure 1.
Superconducting qubits circuit diagrams. The figure describes three types of superconducting qubits and their control mechanisms. (1) Charge qubits consisting of a Josephson junction and a capacitor, with qubit control achieved by adjusting the voltage V_g. (2) Flux qubits which involve a loop inductance (L) and the qubit's energy levels are controlled by varying the bias flux ϕ .(3) Phase qubits which are controlled by adjusting the bias current I, which alters the potential energy surface.

commercial and scientific use. Two other major advantages are the ease of coupling and the relatively simple control mechanisms. The ease of coupling refers to the use of capacitors and inductors, which make it straightforward to link multiple qubits, facilitating the creation of interconnected quantum systems. Additionally, superconducting quantum computers can utilise commercial microwave devices to generate the microwaves necessary for controlling qubits and performing measurements. This not only simplifies the process but also helps reduce the production costs of quantum computers. These advantages underscore why superconducting qubits have become a preferred technology, attracting substantial investment.

The main disadvantage of quantum superconducting circuits is that signals originate from room-temperature equipment, which introduces noise and significantly shortens the coherence time of the qubits. This loss of coherence limits the time available for quantum computations and is a persistent challenge in superconducting quantum circuits. Additionally, both the operation and readout of qubits are highly susceptible to noise and imprecision, further complicating the reliable manipulation of quantum states. This issue remains one of the most significant obstacles in superconducting quantum technology. Another drawback is that superconducting qubits do not exhibit a true two-level quantum transition. Instead, they often interact with higher energy states, leading to potential errors during quantum operations. Therefore, $|0\rangle \rightarrow |1\rangle$ must be carefully avoid in the information processes [25]. In recent years, significant advancements have been made in superconducting quantum computing, particularly in extending coherence time, improving operational fidelity, and enhancing the readout of qubits. One of the most impactful breakthroughs has been the use of optical fibres to link information flow within the device. This innovation has helped reduce noise and signal degradation, contributing to more reliable qubit performance. The implementation of optical fibres represents a major leap forward in addressing key challenges, as will be discussed in this chapter.

2. The use of photo link to improve quantum readout

Until 2021, many of the most promising superconducting quantum computers utilised ensembles with coaxial cables, which were expected to reach a few thousand qubits [26]. The setup (**Figure 2**) includes a microwave modulator that converts the frequency into an electrical signal. This signal travels to another room equipped with cryogenic equipment that reduces the temperature to 20 *mileK*, in a cavity where the

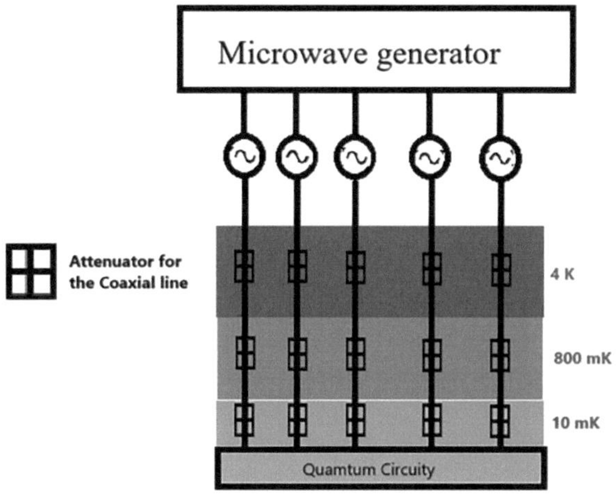

Figure 2.
The standard coaxial superconducting quantum computer involves modulating microwave signals, transforming them into electrical current for transport through a coaxial cable into three distinct vacuum chambers. These chambers maintain temperatures of 4 K, 800 mK, and 10 mK, respectively. The chambers includes an attenuation system crucial for dissipating the heat load generated by active signals the are represented by the black box in the figure. This attenuation is essential to mitigate black body radiation from previous chambers, preventing interference with the delicate quantum circuitry.

quantum circuitry is located. The quantum circuitry manipulates the qubits for logical gates, similar to a classical computer. However, it is necessary to read out the quantum bits to obtain output from the calculations.

The readout of the quantum state of qubits which is used for the calculation in the circuit is done in a three-dimensional cavity coupled to the quantum circuitry, a microwave antenna emits a wave with amplitude and angular frequency α_{in}, ω_d. The frequency ω_l is the transition frequency between the to-measure quantum state. The qubits in the cavity have the coherence state $\alpha_{1,2}$ where the wave has most of the amplitude in the line width of k. Therefore, the electron in the qubits state will absorb and emit another microwave, where the difference between differences in frequency is recorded with a homodyne set-up. The reflected signal has a difference in frequency given by

$$(\omega_d - \omega_l \pm Xik/2)\alpha_{1,2} = i\sqrt{k}\alpha_{in} \tag{2}$$

where X is the difference in phase between the states of the qubits. The amplitude of final wave read out is

$$\alpha_{out} = \sqrt{k}\alpha_{1,2} - \alpha_{in} \tag{3}$$

An increase in temperature introduces challenges related to signal coherence, leading to a reduction in signal duration and impairing accurate readout. Furthermore, the rise in quantum noise becomes a significant issue, as the spectral density of thermal noise becomes indistinguishable from quantum noise [27]. Consequently, elevated temperatures directly exacerbate quantum noise, affecting the system's overall performance

$$H_T(\nu, T) = \left(h\nu/\left(e^{h\nu/k_BT} - 1\right) + (h\nu)/2 \tag{4}$$

Here, H_T is the spectral density of quantum thermal noise that generates a frequency of electromagnetic radiation ν. For the determinate range of values of temperatures T, it creates a high frequency which disturbs the frequency in Eq. (2). Therefore, the temperature has to be taken lower possibly, or the heat load for the coaxial cable became an issue. The coaxial cable has two types of heat [28]: passive heat and active heat. The passive is the flow of heat from the coaxial cable liberated as black body radiation. The heat estimated [26, 28] for 2.2 mm diameter stainless steel cable is $P^{passive}_{coaxial}$=4–13nW. The activated load is joules heated by the attenuation of the signal and is related to the mean power of the electrical pulse in the following equation. $P_{avarege} = \left(\int_0^{\tau}(P(t)/\tau), dt\right.$, where P(t) is the power of the qubit, and the average power is guided by the integration of the time divided by the pulse duration τ. The equation reaches from the measurement of the activated heat and is determined empirical concerning the system

$$P^{active}_{coaxial} = P_{avarege}(1/A - 1) \tag{5}$$

The constant $A < 1$ is given by the composition of the attenuator use in the quantum computer system.

2.1 Improvement of number of qubits using optical fibre

The innovation described in [28, 29] replaces traditional coaxial cables with optical fibres. The new configuration (**Figure 3**) uses multiple lasers transmitted through optical fibres to electro-optic modulators (EOMs) located in a room maintained at 300 K. These EOMs are also connected to a commercial microwave modulator, similar to the previous setup. The EOMs modulate the frequency, phase, and amplitude of the light to match the microwave signals generated by a signal generator. The modulated light, containing all the laser signals, is then transmitted to the cryogenic environment via optical fibres. These fibres exhibit negligible thermal conductivity at low temperatures, where the noise level approaches that of vacuum microwave fluctuations, allowing the generation of highly coherent signals. Additionally, their broad bandwidth enables signal delivery to millions of qubits with minimal passive heat load. The optical carriers are separated by wavelength and directed into high-speed photo-detectors. These detectors utilise advanced techniques to exploit the electron-photon interaction in semiconductor diodes, enabling the rapid conversion of optical information into photocurrent. The resulting electron current mirrors the frequency and phase of the generator and is processed by the circuit, just as in the previous setup. Each incident photon generates an electron-hole pair in the diode with an efficiency of η. The photo-current, which forms a continuous stream, follows the relationships outlined as follows.

$$I = \Re P_0 \tag{6}$$

$$\Re = \eta e/\hbar\omega_{01} \tag{7}$$

I is electron current, P_0 is the incident power of the photon, ω_{01}, e are the frequency of the optical photos and electron charge, and $\hbar$ is the reduced Planck constant. The passive load of the optical fibres is smaller than $P^{passive}_{fibres} = 13pW$ [28, 30], measuring for a silica core of 125 μm more with a coating of the same diameter. The difference in passive heat between coaxial cables and optical fibres is significant, with

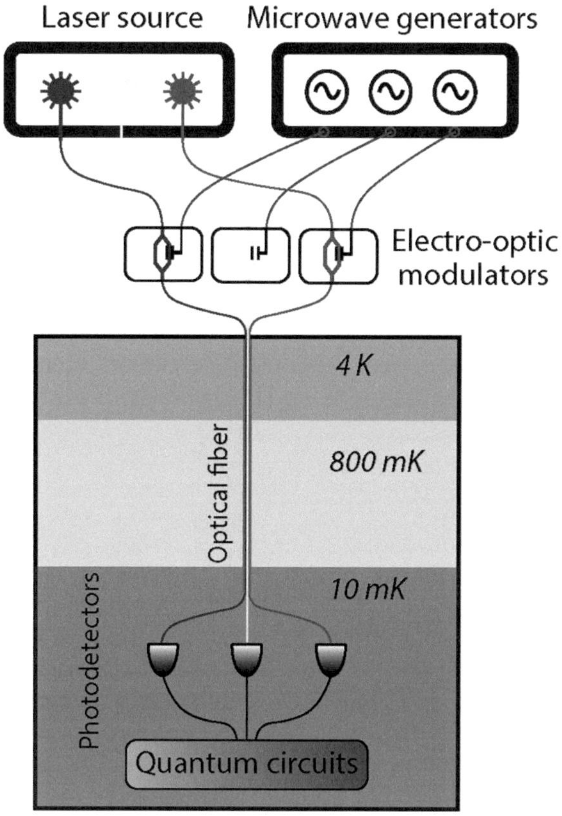

Figure 3.
The fibre optics approach uses a microwave generator signal to modulate laser by using elector-optic modulators (EOMs). The modulation onto an array of optical carriers is transported to the cryogenic room where they are separated by a prism to high-sped photo-detectors. The low thermal conductivity of silica suppresses the passive heat. As for active heat, it can be managed by controlling the independence of the photo-detectors.

optical fibres generating 300–1000 times less passive heat. The active heat load in this approach is proportional to the total energy converted into heat within the photo-detector. However, the researchers did not account for the heat or energy loss between the diode and the control line. The average power dissipated can be expressed as follows:

$$P^{activ}_{fibres} = \sqrt{2P_{avarege}/\Re^2 Z} \tag{8}$$

Here, Z is the impedance of the photodiode. The total heat is given by the sum of the function in relation to the duty cycle.

$$P_{coaxial,fibres} = P^{passive}_{coaxial,fibres} + D^* P^{activ}_{coaxial,fibres} \tag{9}$$

The duty cycle represents the ratio of the qubit charge amplitude or pulse to the qubit's time signal, expressed as a percentage. Consequently, the heat generated is proportional to the power of the wave. If the signal amplitude is relatively small compared to the microwave signal, the resulting heat will also be minimal. When the total power is low, most of the energy is utilised for information transmission. This feature maximises the number of physical qubits, normalised by the cooling power of the mixing chamber, $P_{cooling}$, which is the average medium background power in a temperature at $10mileK$. The number of physical qubits can be determined by dividing the total heat generated by the heat dissipated through the cable or optical fibre

$$N_{qubits} = P_{cooling}/P_{coaxial,fibres} \tag{10}$$

In the fibre optic photon link approach, activation heat can be minimised by selecting lower-frequency lasers and using high-impedance photo-detectors. Additionally, employing smaller duty cycles is essential, though they should not be too small, as this could negatively affect the coherence measurements of the qubits. Quantum computers that use optical fibres support a significantly larger number of qubits, as shown in **Figure 4**.

2.2 The fidelity of qubits

In quantum mechanics, fidelity is defined as the accuracy or procession of the quantum state: the measure how close the determined state is to the target one. For example, for hypothetical two quantum states, a measured one $|\phi\rangle$ and a desire one $|\psi\rangle$.If $(|\phi\rangle/|\psi\rangle)$ = 1, then the two states are the same.On the other hand, if it is 0, they are orthogonal to each other and completely distinct. In quantum computation, the concept is expanded to process fidelity, that measures how closely the quantum operators align with their intended or ideal operation. For the superconducting quantum computer, the fidelity [28, 31] can be calculated by the following equation:

$$F = 1 - P(2|1) - P(1|2) \tag{11}$$

where the probability p(x|y) defines the chances of measuring a qubit in state x when it is prepared in the energy level of the state y. In the absence of information about preparation errors, the fidelity function can be deduced from the error function for a large number of measures for which a Gaussian approach is valid.

$$F = \left(2/\sqrt{\pi}\right) * \int_0^{\sqrt{\eta_1 \Gamma\tau/2}} e^{-t^2} dt \tag{12}$$

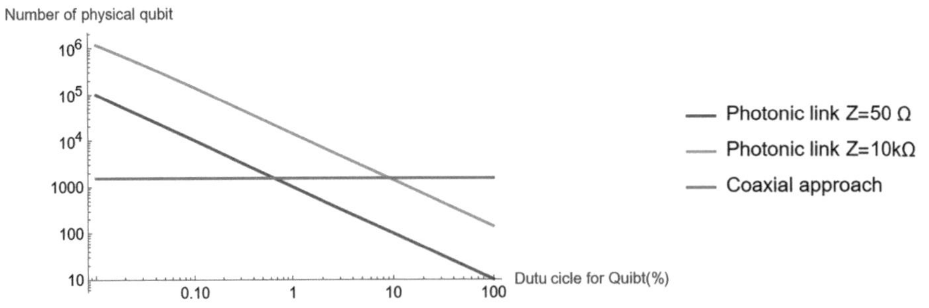

Figure 4.
The maximum number of qubits, as determined by Eq. (10), is influenced by the duty cycle from Eq. (9). For the coaxial cable approach, this number is significantly limited. However, by using optical fibres combined with high-impedance photo-diodes, the maximum number of qubits can be increased by more than 1000 times. The graph was build by data for taken Eq. (10) take form the literature [28].

where τ is the integration time of the measure and η_1 is the efficiency of the measure done by the microwave cavity. Gamma is the distance between two coherent states that is defined by the equation:

$$\Gamma = \sqrt{k}|\alpha_2 - \alpha_1|^2 \tag{13}$$

The parameter γ represents the steady state of a qubit with two different phases. Quantum noise and electron collisions disturb the frequency ω_l and therefore the phase, resulting in an increasing phase difference between the two quantum states. As a result, Eq. (12) is increased. Fidelity is a crucial metric in quantum computation, as low fidelity in quantum gate operations introduces inherent errors. These errors propagate through the system, ultimately affecting the final output. When error propagation becomes significant, obtaining a reliable result is impossible, undermining the integrity of quantum computation. In the systems discussed earlier in this chapter, both achieved a fidelity of 0.98. Notably, the increased number of qubits in the cavity chamber did not compromise the computer's overall fidelity.

3. Additional development of the superconducting quantum computer usage: Optical fibres for better read out

A common issue in the former system is that in the quantum circuity like Josephson-junction circuits operate in the voltage of mV. For the use of a high frequency of 10 Ghz, the rapid single flux quantum or RSFQ. For the readout of the qubits by microwaves, this voltage is not sufficient, so an amplifier is introduced in the quantum circuity to amplify the quantity to hundreds of mV. These amplifiers generate additional heat and constrain the number of physical qubits. Therefore, using another device to replace the cavity was proposed in the following presented works. The first work [32] used a low-backaction electro-optomechanical transducer to read out the qubit into light transmitted by optical fibre. The next work, published in the following year [33], compares the past improvement and original design to new other developed by then using two optical electron transducers, to read out quantum states with great fidelity and small energy. To read out the qubits, the same 3D cavity described in Section 2.1 is used. However, instead of using an antenna for microwave measurements, a laser is directed to an electro-optical transceiver (EO) that converts the light into a microwave signal via optical pumping. This microwave signal enters the cavity to read out the qubits. The return signal then enters the same EO, which reconverts it to an optical signal. This optical signal travels through a fibre back to a room at normal temperature, where it analyzes using heterodyne detection. This method eliminates the need for an antenna, a quantum amplifier, and other microwave components while allowing for better signal analysis via heterodyne detection. The use of optical fibre instead of coaxial cable enables the efficient transmission of qubit data back to the computer room at normal temperature. There are six frequencies of the in-system which can be seen in **Figure 5**. The first ω_c is the readout frequency of the vacuum chamber which is equal to optical free-space frequency, ω_{FSR} refers to the frequency of the transmitted light that exits the optical fibre and reaches the transducer.
The frequency of the electro-optic transducer or ω_e is set in the same value as the previous one [33, 34] for the system to reach the maximum resonance of data possible to achieve the maximum transduction possible. Because if there is a difference, it will cause a delay and loss of information. So, for the measurement or read out of qubit,

$\omega_c = \omega_{FSR} = \omega_e$. The incident laser light to be converted by the electro-optic modulator (EOM) has a frequency of ω_0, while the optical signal from the microwave pump has a frequency of ω_p. The relationship between these two frequencies is given by the following equation:

$$\omega_e = \omega_{FRS} = \omega_0 - \omega_p \tag{14}$$

The last one is the angular velocity of the transition of qubits from state $|e\rangle \rightarrow |g\rangle$, that is, ω_q, where this transformation represents a random change energy, not necessarily the previous referee transformation $|1\rangle \rightarrow |2\rangle$. ω_c will be equivalent to ω_d in Eq. (2). In **Figure 5**, it deployed the diagram of three measurement systems for the qubits. **Figure 5a–c** are the sketches of the distinct of the three systems. The figures a, b, and c represent the three different systems used for the readout of the quantum bit involving a 3D microwave cavity dispersively coupled to a transmon qubit in pink, the electro-optical transceiver EO, consisting of a second microwave cavity blue and grey coupled to an optical whispering gallery mode resonator light and blue. The first is a standard cavity equal to the previous Section 2.1 where the read-out is performers from a microwave antenna, and the data are transmitted by a coaxial cable. The second is the use of EO and laser to convert the signal for microwave, but the transmission is made by coaxial cable. The last is the system discussed in this section. The display of the skeet of their equipment is in **Figure 5**, the rectangular green boxes

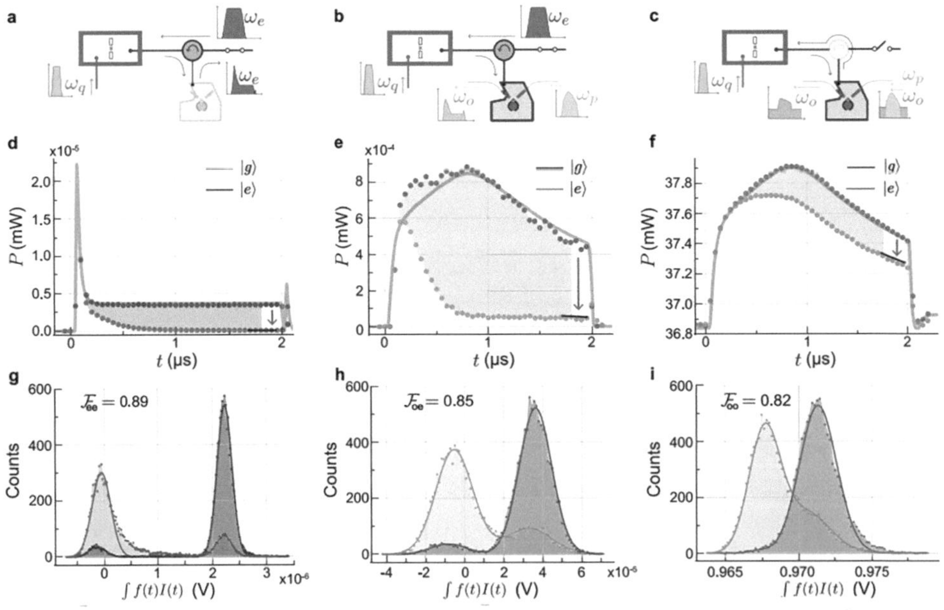

Figure 5.
(a) Conventional microwave readout: a microwave pulse probes the cQED system and is detected via microwave heterodyne detection. (b) The microwave pulse reflected from the cQED system is upconverted to the optical domain and detected with optical heterodyne detection. (c) A modulated optical signal is converted to the microwave domain to probe the cQED system. Its reflection is simultaneously converted back to the optical domain and detected with an optical heterodyne setup. The graphs d,e, and f represent the average energy of the signal. Finally, the graphs g,h, and i represent the number of all signs added or counts signal for the time weight function $f(t) * (I(t)$ multiplied by the in-phase quadrature of microwave output field or total field of an electromagnetic wave passing a volume $I(t)$. Finally, it displays the fidelity of each system for arbitrary measures of the quantum state (figure adapted from reference [33] have free licences CCBV 4.0).

DOI: http://dx.doi.org/10.5772/intechopen.1007572

represent the circuits are the transitions are happening, the incomplete rectangle in blue is the EO, and the purple circle is the 3D cavity.

The graphs displayed in (d–f) represent the output photons or microwaves measured by the heterodyne signal powers. It is evident that the power increases by a factor of 40 from the system (d) to (e), 40,000 from (e) to (f), and more than 1,600,000 from (d) to (f). This significant improvement demonstrates the potential to reduce the initial input power of the signal generated by the microwave generator, which could lead to more power-efficient computers. Consequently, by simplifying the electronics and reducing power consumption, new, more compact computers can be developed. These advancements pave the way for scaling down devices to fit small spaces such as satellites and research stations. Additionally, there is a minor decrease in fidelity values displayed in (g), (h), and (i). This can be attributed to the overlap of signals from the two states, which generates fidelity errors in the measurements. Although more experiments and significant results have been achieved in this work [33], they will not be covered in this chapter due to space constraints. Therefore, this new system has proven capable of replacing the cavity microwave and cryogenic amplifier in computer circuitry, significantly reducing the electronic cryogenics component. The new features available open the possibility of reducing the size of superconducting quantum computers, making them smaller and more energy-efficient. Moreover, the use of this technology in space, where temperatures approach $0K$, makes it feasible to design quantum super-conductors in satellites, potentially giving rise to a future quantum global network.

3.1 Other relevant improvement of the previous system

Another relevant article [35] using piezo-optomechanical transducer achieved a readout with fidelity of 0.99 at 6 μW, each was better than 0.98 of 2021 article [28], with the same feature of the previous one eliminating the use of cryogenics microwave amplifiers. Using a chip of superconducting electro-optic modulator [36] or SEOM, that is cable of read out the qubits states. This has a great bandwidth of 17 GHz that has read out at a voltage level of 42 mW that can dispense any amplifier. Furthermore, this great bandwidth is a cable of direct data lifting from the quantum circuity via optical fibre as shown in **Figure 6**. The SEOM is considered by the author [36] an important step to build a high-bandwidth cryogenic integrated circuity with optical links. The complete explanation of these articles in the chapter breaches the

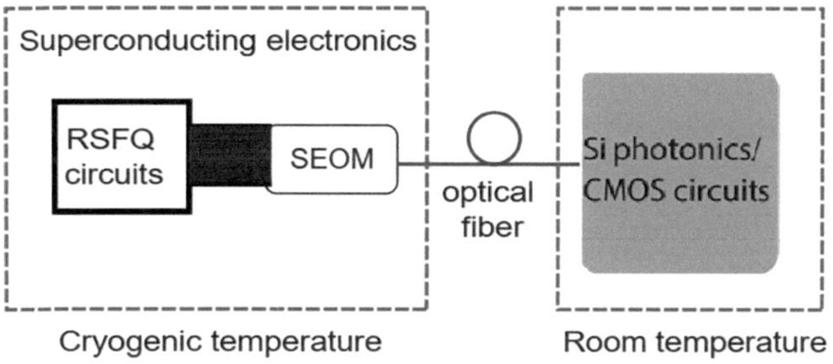

Figure 6.
The diagram of the superconducting quantum circuitry, based on the Rapid Single Flux Quantum (RSFQ) technology, is connected to a superconducting electro-optic modulator (SEOM) in a cryogenic environment. This setup is integrated with silicon photonic CMOS circuitry for data acquisition and processing on a computer. The connection is facilitated through a single optical fibre.

permitted scope. The authors in their final remarks emphasise the importance of the optical fibre together with optical-electro modulators and traducer cable for further improvement of the superconducting computers.

4. Conclusion

The use of optical fibres has once again proven indispensable for the future of quantum computer hardware. They are expected to significantly enhance the feasibility of quantum computers with millions of qubits, enabling more compact setups. This advancement paves the way for the development of next-generation superconducting quantum computers, despite the many technological and physical challenges that still need to be addressed.

Acknowledgements

I would like to express my gratitude to Maja Bozicevic, the Publishing Process Manager at IntechOpen, for her unwavering support. Her reminders about my chapter and its deadlines, as well as her assistance throughout the process, have been invaluable. Additionally, I am deeply thankful to her for providing me with the incredible opportunity to contribute a chapter to this book.

Author details

Ricardo Rovere de Santi
LabM^2 at IF-USP, University of Sao Paulo, Sao Paulo, Brazil

*Address all correspondence to: ricardo.santi@alumni.usp.br

References

[1] Benioff P. The computer as a physical system: A microscopic quantum mechanical Hamiltonian model of computers as represented by turing machines. Journal of Statistical Physics. 1980;**22**:563-591

[2] Feynman RP. Simulating physics with computers. International Journal of Theoretical Physics. 1982;**21**(6/7):467-488

[3] Manin YI. Vychislimoe i Nevychislimoe (Computable and Noncomputable). Moscow: Sov; 1980

[4] Kalinowski M, Maskara N, Lukin MD. Non-abelian floquet spin liquids in a digital rydberg simulator. Physical Review X. 2023;**13**(3):031008

[5] Choi J et al. Preparing random states and benchmarking with many- body quantum chaos. Nature. 2023;**613**(7944): 468-473

[6] Lee JY et al. Landau-forbidden quantum criticality in rydberg quantum simulators. Physical Review Letters. 2023;**131**(8):083601

[7] Shi Y-H et al. Quantum simulation of topological zero modes on a 41-Qubit superconducting processor. Physical Review Letters. 2023;**131**:080401. DOI: 10.1103/PhysRevLett.131.080401

[8] Fernandez-Carames TM, Fraga-Lamas P. Towards post-quantum blockchain: A review on blockchain cryptography resistant to quantum computing attacks. IEEE Access. 2020;**8**: 21091-21116

[9] David P. DiVincenzo. "The physical implementation of quantum computation". Fortschritte der Physik: Progress of Physics. 2000;**48**(9–11): 771-783

[10] Clarke J, Wilhelm FK. Superconducting quantum bits. Nature. 2008;**453**(7198):1031-1042

[11] Krantz P et al. A quantum engineer's guide to superconducting qubits. Applied Physics Reviews. 2019;**6**(2): 021318-1-021318-57

[12] Cirac JI, Zoller P. Quantum computations with cold trapped ions. Physical Review Letters. 1995;**74**(20): 4091

[13] Bruzewicz CD et al. Trapped-ion quantum computing: Progress and challenges. Applied Physics Reviews. 2019;**6**(2):021314-1-021314-56

[14] Urban E et al. Observation of Rydberg blockade between two atoms. Nature Physics. 2009;**5**(2):110-114

[15] Xiaoling W et al. A concise review of Rydberg atom based quantum computation and quantum simulation. Chinese Physics B. 2021;**30**(2):020305

[16] DeMille D. Quantum computation with trapped polar molecules. Physical Review Letters. 2002;**88**(6):067901

[17] Kane BE. A silicon-based nuclear spin quantum computer. Nature. 1998; **393**(6681):133-137

[18] Burkard G et al. Semiconductor spin qubits. Reviews of Modern Physics. 2023;**95**(2):025003

[19] Cory DG, Fahmy AF, Havel TF. Ensemble quantum computing by NMR spectroscopy. National Academy of Sciences of the United States of America. 1997;**94**(5):1634-1639

[20] Jones JA. Quantum computing with NMR. Progress in Nuclear Magnetic

Resonance Spectroscopy. 2011;**59**(2):91-120

[21] Jeremy L, O'brien. Optical quantum computing. Science. 2007;**318**(5856): 1567-1570

[22] Wang J et al. Integrated photonic quantum technologies. Nature Photonics. 2020;**14**(5):273-284

[23] Shen LYL. Superconductivity of tantalum, niobium and lanthanum studied by electron tunneling: Problems of surface contamination. pp 31-44 of superconductivity in d- and f-band metals. /Douglass, D. H. (ed.). New York American Inst. of Physics (1972). Other information: From conference on superconductivity in d- and f-band metals; Rochester, N. Y. (29 Oct 1971). Orig. Receipt Date: 31-Dec-72. In: AIP Conference Proceedings. Vol. 4.1. American Institute of Physics; 1972

[24] Kjaergaard M et al. Superconducting qubits: Current state of play. Annual Review of Condensed Matter Physics. 2020;**11**:369-395

[25] Huang H-L et al. Superconducting quantum computing: A review. SCIENCE CHINA Information Sciences. 2020;**63**(8):180501-1-180501-32. ISSN: 1869-1919. DOI: 10.1007/s11432-020-2881-9

[26] Krinner S et al. Engineering cryogenic setups for 100-qubit scale superconducting circuit systems. EPJ Quantum Technology. 2019;**6**(1):2

[27] Kerr AR, Feldman MJ, Pan S-K. Receiver noise temperature, the quantum noise limit, and the role of the zero-point fluctuations. In: Blundell R, Tong E, editors. Proceedings of the 8th International Symposium on Space Terahertz Technology. 1997, Proceedings of a Conference Held in Cambridge, MA; 25-27 March 1997. Sponsored by JPL Center for Space Microelectronics Technology and Harvard University; 1997. pp. 101-111. Originally printed as Electronics Division Internal Report No. 304, National Radio Astronomy Observatory, Charlottesville VA 22903, September 1996

[28] Lecocq F et al. Control and readout of a superconducting qubit using a photonic link. Nature. 2021;**591**(7851): 575-579

[29] Youssefi A et al. A cryogenic electro-optic interconnect for superconducting devices. Nature Electronics. 2021;**4**(5): 326-332

[30] Smith TL, Anthony PJ, Anderson AC. Effect of neutron irradiation on the density of low-energy excitations in vitreous silica. Physical Review B. 1978; **17**:4997-5008. DOI: 10.1103/PhysRevB.17.4997

[31] Houck AA et al. Controlling the spontaneous emission of a superconducting transmon qubit. Physical Review Letters. 2008;**101**: 080502. DOI: 10.1103/PhysRevLett.101.080502

[32] Delaney RD et al. Superconducting-qubit readout via low-backaction electro-optic transduction. Nature. 2022; **606**(7914):489-493

[33] Arnold G et al. All-optical single-shot readout of a superconducting qubit. 2023:2310.16817-1-2310.16817-15. Available from: https://arxiv.org/abs/2310.16817

[34] Wallraff A et al. Approaching unit visibility for control of a superconducting qubit with dispersive readout. Physical Review Letters. 2005;**95**: 060501. DOI: 10.1103/PhysRevLett.95.060501

[35] van Thiel TC et al. Optical readout of a superconducting qubit using a scalable piezo-optomechanical transducer. 2023. 2310.06026-1-2310.06026-15. Available from: https://arxiv.org/abs/2310.06026

[36] Shen M et al. Photonic link from single-flux-quantum circuits to room temperature. Nature Photonics. 2024;**18** (4):371-378

Chapter 7

Different Modulation Schemes Employed in Free Space Optical Communication Systems

Bipin Kumar Saw, Ghanshyam Singh and Manish Tiwari

Abstract

Numerous modulation schemes are used in free space optical (FSO) transmission, including Phase Shift Keying (PSK), Orthogonal Frequency Division Multiplexing (OFDM), Space Shift Keying (SSK), Pulse Position Modulation (PPM), Gaussian Minimum Shift Keying (GMSK), and Subcarrier Intensity Modulation (SIM). By changing the phase of digital data, PSK changes the spectrum into orthogonal subcarriers, OFDM changes the beam's location, PPM changes the positions of pulses, GMSK uses a Gaussian filter to smooth phase shifts, and SIM changes the amplitude of subcarriers. Each scheme is designed to optimally handle different transmission situations, balancing data rate and reliability based on factors like system complexity, available bandwidth, and the state of the atmosphere.

Keywords: free space optical (FSO), intensity modulation/direct detection (IM/DD), pulse modulation, phase modulation, orthogonal frequency division multiplexing (OFDM), Gaussian minimum shift keying (GMSK), subcarrier intensity modulation (SIM)

1. Introduction

Free space optical (FSO) communication is an emerging technology that employs light as a carrier for transmitting information and has gained significant attention due to its potential for high-speed, secure, and license-free communication [1]. These characteristics make FSO a preferred option for last-mile access, connecting end-users to existing fiber optic networks and an important component in the development of future wireless data networks, presenting significant opportunities for integration into fifth-generation (5G) wireless systems. With the increasing need for data worldwide, especially due to the growth of 5G, the Internet of Things (IoT), and data-intensive applications, FSO technology offers an appealing substitute for reducing the congestion that arises in radio frequency (RF) systems. Also, it is a viable option for tying up local area networks (LANs), wide area networks (WANs), and metropolitan area networks (MANs), as a result of the expansion of communication networks. Even if light-based information can be sent by fiber cables with efficiency, there are situations in which the free space between the transmitter and the receiver is the only means of communication. Furthermore, compared to fiber optics, FSO systems offer deployment efficiency in terms of both cost and time.

A direct line-of-sight (LOS) connection between the transmitter and the receiver is necessary for FSO communication. It employs a modulated laser beam to create a communication link across the atmosphere and can transfer data at up to gigabit speeds [1–3]. While the advantages are evident, FSO communication systems are not without challenges. The laser beam experiences signal distortions due to turbulence, absorption, and scattering as it travels through the atmosphere, which affects the quality of the signal that is received which gives a bandwidth limitation. Secure data communication is also a primary concern for today's communication. Even optical beams are highly directional and confined to the line-of-sight which makes eavesdropping significantly more difficult, but due to beam spreading, there is a chance for eavesdropping. Therefore, there is a need for high secure FSO communication by limitation of beam spreading.

2. FSO communication system overview

In the initial step of an FSO communication system, a modulated narrow beam of light is emitted from a laser source at the transmitter end. This modulated light beam then traverses through the free space until it reaches the receiver. A simple block diagram of an FSO communication system with two sets of transceivers for a full-duplex link is shown in **Figure 1**.

The information is encoded onto the optical carrier's amplitude, frequency, phase, or intensity to facilitate communication. Subsequently, the modulated optical signal is transmitted through the free space.

A lens system collects the incident optical field at the receiver, and then a photodetector converts it into an electrical signal. This electrical signal undergoes further processing to recover the transmitted information.

3. Challenges in FSO communication system

There are a lot of challenges associated with FSO in addition to its benefits. These systems depend a lot on the atmospheric conditions which introduce errors in the system. In FSO link, the attenuation primarily results from factors such as absorption due to water vapor, scattering caused by water droplets, and atmospheric turbulence which includes scintillation and beam spreading. Among these, atmospheric turbulence is a major obstacle leading to severe degradation in FSO link performance. Thus,

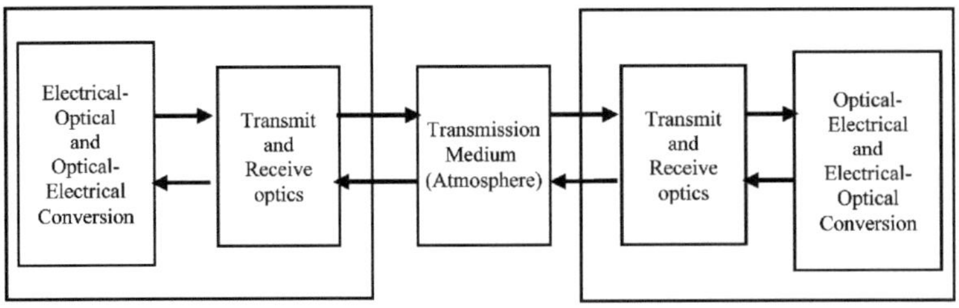

Figure 1.
Bidirectional free space optical communication system.

the atmosphere degrades the signal in FSO links through various mechanisms, including absorption, scattering, and scintillation.

Another contributing factor to attenuation in FSO links is geometric attenuation loss. Proper alignment of the transmitted light beam with the transceiver's receiving aperture at both ends is essential for an FSO link to operate properly. However, this alignment can be disrupted by factors such as wind effects, small earthquakes, and vibrations from transmitter and receiver platforms. Misalignment prevents the establishment of a line-of-sight link, resulting in heavily attenuated or sometimes entirely lost received power. Thus, this pointing error (PE) should also be considered in the loss of the FSO link.

4. Channel models in FSO communication system

In the form of a probability density function (PDF), several models have been developed to simulate the irradiance of the atmospheric channel from weak to strong regimes. Examples of such PDFs are Lognormal [4, 5], Gamma-Gamma [5, 6], K-distribution [7], extended Generalized-K [8], double Weibull [9], I-K distribution [10], Malaga ($\mathcal{M}$)-Distribution [11], inverse Gaussian-Gamma distribution [12], and Fisher-Snedecor distribution ($\mathcal{M}$) distribution [13]. In these models, Gamma-Gamma distribution model has been widely adopted by researchers. This is because it is a good fit for the statistics of the fading and scintillation effects that occur due to atmospheric turbulence (AT). Apart from Gamma-Gamma distribution, $\mathcal{M}$ distribution is also famous among researchers. This is because the PDF of the $\mathcal{M}$ distribution is a generalized PDF, i.e. most existing PDFs are the special cases of $\mathcal{M}$ distribution. Its PDF is derived by multiplying the shaded Rice and Gamma distributions. It is proposed in Ref. [11] and the performance of $\mathcal{M}$ distribution AT with PE is analyzed in Ref. [14]. In addition to Fisher-Snedecor ($\mathcal{F}$) distribution, authors in Ref. [13,15] have introduced $\mathcal{F}$ distribution for FSO communication system and they have mentioned that this $\mathcal{F}$ distribution is the best exact fit with the experimental data. The PDF and cumulative density function (CDF) expressions are also simpler than the $\mathcal{M}$ distribution, Gamma-Gamma, and other distributions. In $\mathcal{F}$ distribution, small-scale irradiance function is modeled by Gamma distribution and large-scale irradiance function is modeled by inverse Gamma distribution. The PDF and CDF have derived for $\mathcal{F}$ distribution with PE in Ref. [16] and also derived the average bit error rate (ABER) expression for different amplitude/phase modulation (APM) schemes.

5. Modulation schemes in FSO communication system

In order to provide reliable and effective communication, modulation schemes play a vital role. It is a process in which a message is coded onto the optical carrier signal by altering specific characteristics of the optical signal, such as amplitude, phase, frequency, or polarization to represent the desired information. System requirements, channel conditions, available bandwidth, and desired data rate are only a few of the factors that influence the modulation scheme choice. Selecting the appropriate modulation scheme in FSO communication requires careful consideration of trade-offs, including data rate, bandwidth efficiency, link distance, atmospheric conditions, system requirements, and receiver complexity. A few of the modulation schemes will be discussed in the following section.

5.1 IM/DD FSO systems

The IM/DD modulation scheme is widely used in FSO communication systems due to its simple implementation. In this modulation scheme, direct modulation is used on the transmitter side by encoding the information onto the intensity of the optical signal. Then, the transmitted optical signal is directly projected onto the photodetector through the lens system of the receiver. The received instantaneous intensity under the assumption of a turbulence-free channel can be written as [17]:

$$I(t) = I_r(1 + km(t)) \tag{1}$$

where, I_r is the average received intensity, $m(t)$ is the modulating signal, and k is a scaling factor. Here, transmitting intensity is always positive, i.e. $|km(t)| \leq 1$.

5.1.1 OOK

On–Off Keying (OOK) is a type of IM/DD modulation in which binary data is represented by the presence or absence of the optical carrier signal. OOK is a mostly used modulation scheme for IM/DD modulation in practical FSO systems. It is a simple and efficient method for encoding information onto an optical carrier wave. OOK modulation is often employed when simplicity and power efficiency are priorities, such as in FSO links with low data rates.

A logical '1' is conveyed by turning on the carrier wave, while a logical '0' is represented by turning off the carrier. This on-off switching is synchronized with the data to be transmitted, allowing the receiver to extract the original information. The simplicity of OOK modulation makes it easy to implement and suitable for low-complexity transceiver designs.

In terms of implementation, OOK modulation can be realized using different pulse shapes. Two common pulse formats used in OOK are Return-to-Zero (RZ) and Non-Return-to-Zero (NRZ). In RZ-OOK, the carrier is active only for a fraction of the bit period. A logical '1' is represented by a pulse that occupies a fraction of the bit period and returns to zero before the next bit begins. For a logical '0', the carrier remains off during the entire bit period. The RZ-OOK pulse shape ensures better synchronization and facilitates clock recovery at the receiver.

In NRZ-OOK, the optical carrier is held constant for the entire duration of a bit period when transmitting a logical '1'. When transmitting a logical '0', the carrier is completely turned off. This results in rectangular pulses with a constant amplitude.

The choice between NRZ-OOK and RZ-OOK depends on various factors, including the specific application, desired spectral efficiency, noise resilience, and receiver design. OOK-NRZ and OOK-RZ with a duty cycle (λ_{dc}) of 0.5 are shown in **Figure 2** for an average transmitted power (P_{avg}). The envelope for OOK-NRZ can be written as [17]:

$$P(t) = \begin{cases} 2P_r \text{ for } t\epsilon[0, T_b) \\ 0 \textit{ elsewhere} \end{cases} \tag{2}$$

where P_r is the average power and T_b is the bit duration.

The OOK-NRZ modulation scheme achieves a power efficiency (η_p) of tw and a bandwidth efficiency (η_B) of one. However, the bandwidth efficiency of the OOK-RZ scheme is dependent on the duty cycle, yet it displays the same power efficiency as the

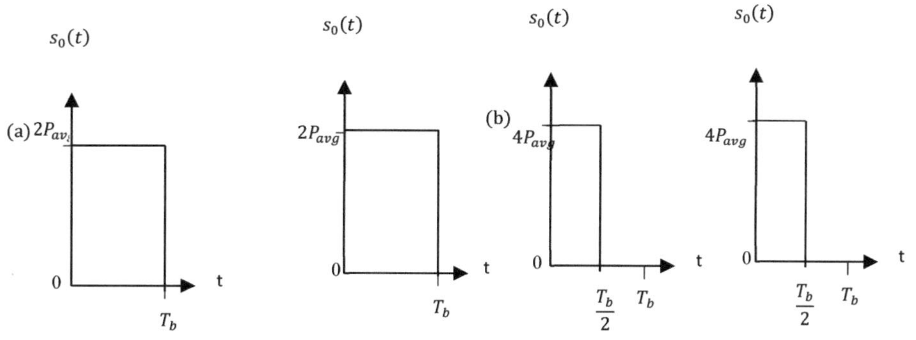

Figure 2.
OOK transmitted waveform (a) NRZ (b) RZ ($\lambda = 0.5$).

OOK-NRZ scheme. For no turbulence, the conditional bit error rate (P_{e-OOK}) for NRZ-coded optical data can be represented as a function of SNR as follows [18]:

$$P_{e-OOK} = \frac{1}{2}\text{erfc}\left(\frac{1}{2\sqrt{2}}\sqrt{SNR}\right) \tag{3}$$

Numerous efforts have been done by researchers for adapting and advancing the OOK modulation scheme. In Ref. [19], X. Zhu et al. demonstrated that employing maximum-likelihood sequence detection (MLSD) and pilot-symbol-assisted detection (PSAD) at the receiver results in signal-to-noise ratio (SNR) gains of 2.4 and 1.4 dB, respectively, for a link length of 500 m in FSO communication using the OOK modulation scheme. In another work by A. J. Navas et al. [20], here authors derived expressions for the ABER for OOK modulation under the $\mathcal{M}$ distribution, considering factors such as aperture size, beam width, jitter variance, and PE.

5.1.2 SIM

Another modulation technique used in IM/DD systems is SIM [21], which is proposed by W. Huang et al. in Ref. [21] for FSO communication systems. In this approach, the information signal is pre-modulated into the RF electrical subcarrier signal. Various modulation schemes, such as BPSK, QPSK, QAM, amplitude modulation (AM), frequency modulation (FM), etc., can be applied to modulate the electrical subcarrier. Next, the intensity of the optical carrier is modulated using this pre-modulated signal. Similar to IM/DD systems, the signal is recovered using direct detection at the receiver end. Unlike OOK schemes, SIM does not necessitate an adaptive threshold and provides enhanced bandwidth efficiency compared to PPM [21]. Optical SIM leverages the well-established RF system, simplifying the implementation process. Due to this technique, it is possible to transmit multiple information signals simultaneously over the optical link. In **Figure 3**, the principle of the SIM optical system is shown for an FSO link [17, 22]. To achieve subcarrier multiplexing, frequency division multiplexing (FDM) is used to combine different modulated subcarrier signals. After that the intensity of a continuous laser source is modulated using subcarrier multiplexed signal. However, a drawback of this multiplexing scheme is the requirement for precise synchronization and the complexity of the design at the receiver side.

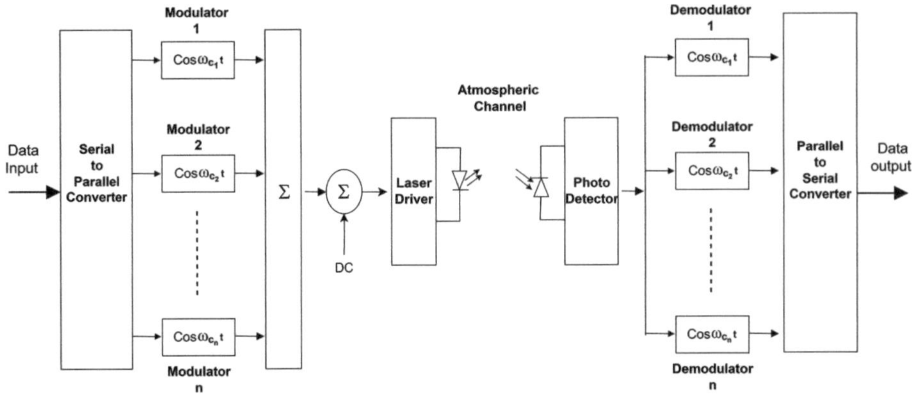

Figure 3.
Subcarrier intensity modulation block diagram.

The main drawback of SIM with IM/DD lies in its suboptimal optical average power efficiency. This is because there are positive and negative values in the SIM electrical signal, and both must be taken into account. To ensure that $x(t)$ remains non-negative, a DC offset must be added. The lowest value of the SIM signal decreases and becomes more negative as the number of subcarrier signals increases. As a result, the required DC bias increases which leads to a decline in optical power efficiency, since the average optical power is related to the DC bias. Because an increase in subcarriers results in an increase in transmitted power, which is unsafe for the eyes, the number of subcarriers is further restricted due to these concerns.

5.2 Pulse modulation

Pulse modulation in FSO communication is a method used to transmit data over optical links by adjusting the intensity of light pulses. FSO communication relies on laser beams sent through the atmosphere to establish fast wireless connections. Pulse modulation is vital for shaping these light signals to effectively carry information. Various pulse modulation techniques, such as Pulse Amplitude Modulation (PAM) and PPM, are employed in FSO communication. Each technique has its own advantages and is suitable for different situations, considering factors like transmission distance, data rate, and atmospheric conditions.

In FSO communication, pulse modulation techniques are crucial for minimizing the impact of atmospheric conditions like fog, rain, and atmospheric turbulence. These conditions can degrade the signal and introduce errors in the received data. Additionally, pulse modulation techniques in FSO communication enable high data rates and are instrumental in establishing dependable and efficient wireless communication. They also compensate for the effects of atmospheric conditions, ensuring a high level of link performance.

5.2.1 PAM

PAM plays a vital role in FSO communication by modulating data onto optical signals by varying the amplitude of light pulses. It provides simplicity, efficiency, and adaptability for wireless information transmission. By using a set of M symbols

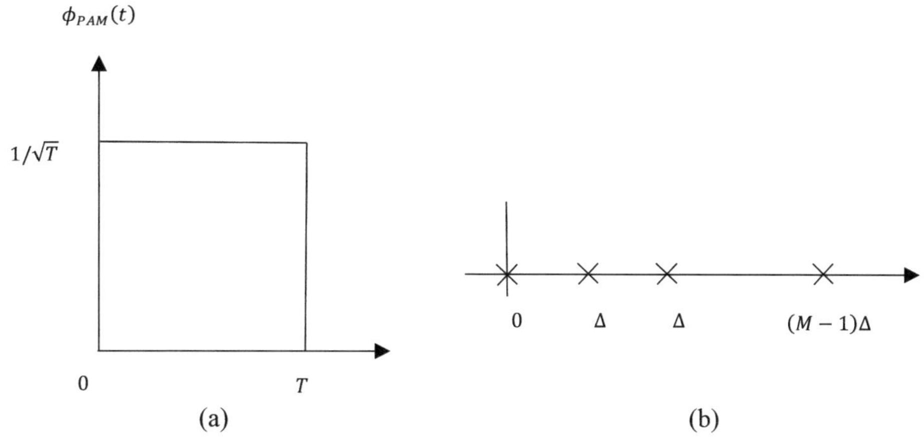

Figure 4.
(a) PAM basis function (b) PAM constellation where $\Delta = 2\sqrt{TP/L} - 1$.

instead of just two, this modulation scheme extends the idea of OOK. **Figure 4(a)** shows a rectangular M-PAM signal defined by the basis function.

Prior knowledge of the data distribution is required in PAM so that the amplitude of the average output remains fixed at P, ensuring the control of average optical power, whereas in PPM, the chosen symbols do not affect the average power of the modulation scheme. In order to comply with the constraint of optical intensity being non-negative, M non-negative scaling factors are chosen. To satisfy the channel constraint, the time-dependent PAM signal can be expressed as

$$x(t) = \sum_{k=-\infty}^{\infty} \frac{2P}{M-1} \sqrt{T}(A[k] - 1)\phi_{PAM}(t - kT) \tag{4}$$

where $A[k]$ is uniform over M. Here, for $M = 2$, this formula corresponds to the OOK intensity waveform.

The constellation diagram of this modulation scheme is one-dimensional, since a single basis function may be used to represent all of the signal points. The distribution of the M constellation points along the axis is uniform as shown in **Figure 4(b)**. Assuming an equal probability of transmission for each symbol, the symbol error probability is given as

$$P_{e_{sym}} = \frac{2}{M}(M - 1)Q\left(\frac{P}{M-1}\sqrt{\frac{1}{R_s\sigma^2}}\right) \tag{5}$$

Furthermore, in FSO communication, PAM can be combined with other modulation techniques like PPM to improve the system performance. Through the utilization of hybrid modulation schemes, FSO communication systems can achieve higher data rates and increased resilience against atmospheric impairments, thereby improving the reliability and efficiency of the transmission.

5.2.2 PPM

This modulation scheme belongs to the family of pulse modulation and is considered an orthogonal modulation scheme. In comparison to OOK, PPM provides better power efficiency, but at the expense of increased complexity and bandwidth requirements. In PPM, each block of $\log_2 M$ data bits is assigned to one of M possible symbols, denoted by the order M-PPM. Each symbol consists of a pulse with a constant power, P_T, occupying one slot, while the remaining $M - 1$ slots are left empty. The information is encoded according to the position of the pulse within the symbol, which translates to the decimal value of the $\log_2 M$ data bits.

The following expression describes the relationship between the duration of each bit and the duration of each slot, T_s:

$$T_s = \frac{T \log_2 M}{M} \tag{6}$$

In **Figure 5**, the transmitted waveforms for 16-PPM and OOK are shown.

For demodulation, a PPM receiver requires synchronization at both the slot and symbol levels. Despite this requirement, PPM remains an effective modulation scheme for FSO communication systems due to its superior power efficiency.

In this modulation scheme, each pulse emitted by a laser represents one or more bits of information based on its temporal position within a symbol. The symbol duration matches that of the information bits. Rather than transmitting bits individually, the encoding is performed in blocks. For optical block encoding, each M bit word is transformed into one of $L = 2M$ optical fields for transmission.

The bit rate is expressed as follows:

$$R_b = B \frac{\log_2 L}{L} \tag{7}$$

The expression of BER for PPM scheme is given as:

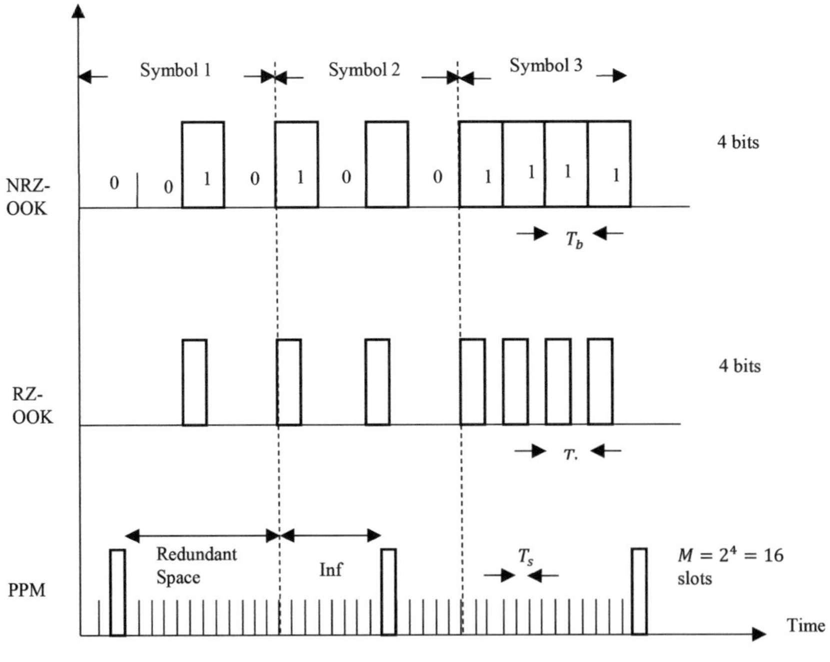

Figure 5.
NRZ-OOK, RZ-OOK, and PPM line coding.

$$P_{b_{PPM}} = \frac{1}{2}\operatorname{erfc}\left(\sqrt{SNR}\right) \tag{8}$$

Dynamic thresholding is not required for optimal detection in PPM when using hard signal detection at the receiver [23, 24]. PPM is very much popular in deep space communication [25]. K. Kiasaleh [26] has analyzed the PPM modulation under Log-Normal and Negative Exponential distribution model while considering the thermal noise for the avalanche photodiode (APD)-based receiver. In Ref. [27], Xiang Yi et al. have derived the ABER expression for M-*ary* PPM (MPPM) modulation scheme over Gamma-Gamma turbulence model.

5.3 Phase modulation

Phase Modulation (PM) is a modulation technique that involves transmitting data by altering the instantaneous phase of an alternating carrier wave. This modulation scheme is applicable to both analog and digital data transmission.

In analog PM, the carrier signal's phase experiences continuous variations. The degree of the carrier-phase shift at any given moment is directly proportional to the magnitude of the positive or negative amplitude of the input signal.

On the other hand, in digital PM, the phase of the carrier signal changes instantaneously instead of undergoing continuous fluctuations. When there are two possible phase states, it is known as bi-phase modulation. More complex modes may involve four, eight, or even more distinct phase states, with each specific digital input data state represented by a corresponding phase angle.

5.3.1 BPSK

The modulation scheme is termed BPSK, when a binary digital signal is utilized to manipulate the phase of a sinusoidal carrier. Depending on the value of the modulating signal, the carrier's phase is set to either 0 or π. When transmitting a '1' symbol, the modulated signal aligns with the carrier phase of 0 while a phase of π represents a '0' symbol.

Here, each modulating signal corresponds to a transmitted symbol with a duration of T_s, which is equal to the bit duration T_b. As a result, the bandwidth is equal to the bit rate in BPSK which is the same as OOK. Then, the bandwidth efficiency of BPSK can be defined as:

$$\eta_{BPSK} = \frac{R_b}{B} = 1 \tag{9}$$

where B is the equivalent bandwidth and R_b is the bit rate. In other words, the theoretical bandwidth efficiency of the BPSK is unit. The conditional BER of BPSK is given as [17]:

$$P_{b_{BPSK}} = \frac{1}{2}\operatorname{erfc}\left(\sqrt{SNR}\right) \tag{10}$$

The normalized power requirement comparison between BPSK and NRZ-OOK can be expressed as follows:

$$\frac{P_{BPSK}}{P_{OOK}} = \frac{1}{2\sqrt{2}} \tag{11}$$

In theory, to attain specific BER performance, the NRZ-OOK could need up to twice as much power as BPSK.

5.3.2 DPSK

In DPSK, no phase difference is represented by '0' and a π phase difference is represented by '1', when compared to the previous phase. The receiver of DPSK is equipped with the capability of storage which is used to determine the relative phase difference between the waveforms that are received across two consecutive bit intervals. DPSK is an example of non-coherent orthogonal modulation and it resolves the issue of inverted π.

In DPSK, unlike in BPSK, synchronous demodulation does not require knowledge of the carrier signal's phase and frequency. However, demodulation still requires a local carrier. The BER expression of DPSK is given as:

$$P_{b_{DPSK}} = \frac{1}{2}\,\mathrm{erfc}\left(\frac{\sqrt{\mathrm{SNR}}}{\sqrt{2}}\right) \tag{12}$$

In DPSK, the required bandwidth is equivalent to the bit rate and it is given by

$$B_{DPSK} = R_b \tag{13}$$

DPSK exhibits a comparatively higher spectrum efficiency and offers improved tolerance toward dispersion, non-linearity, and polarization mode dispersion (PMD). The receiver sensitivity of DPSK is 3 dB higher than that of OOK modulation for the same BER. Also, DPSK performs better than OOK in terms of noise resistance. The bandwidth efficiency of DPSK is equivalent to that of BPSK and OOK:

$$\eta_{DPSK} = 1 \tag{14}$$

The ratio of average power requirement of DPSK to OOK is given as [17]:

$$\frac{P_{DPSK}}{P_{OOK}} = \frac{1}{2\sqrt{2}} \tag{15}$$

5.3.3 QPSK

QPSK, a modulation technique formed by combining two orthogonal BPSK signals, exhibits four distinct phases during transmission. This configuration allows for improved spectral efficiency. As a result, each individual bit occupies a time duration of T_b seconds, and the signals representing the bits last for $T_s = 2T_b$ seconds. As a result, BPSK modulation requires half the bandwidth that QPSK modulation requires. Theoretically, the bandwidth efficiency of QPSK is double that of BPSK:

$$\eta_{QPSK} = \frac{R_b}{0.5R_b} = 2 \tag{16}$$

However, in practical terms, the bandwidth efficiency of QPSK typically ranges between 1.4 and 1.6 bits per second per hertz (bps/Hz). The BER for QPSK can be expressed as:

$$P_{b_{QPSK}} = \mathrm{erfc}\left(\sqrt{\mathrm{SNR}}\right) = 2P_{b_{PSK}} \tag{17}$$

Thus, the following expression can be used to express the average power required using QPSK normalized to NRZ-OOK:

$$\frac{P_{QPSK}}{P_{OOK}} = \frac{1}{2\sqrt{2}} \frac{\mathrm{erfc}^{-1}(\mathrm{BER})}{\mathrm{erfc}^{-1}(2\mathrm{BER})} \tag{18}$$

Hence, the average power demand for QPSK is nearly identical to that of BPSK, with only a slight increase in power consumption.

In Ref. [28], X. Song et al. analyzed the M- *ary* MPSK and M- *ary* MDPSK and obtained the ABER expressions for Log-Normal turbulence model. In this continuation, R. Singh et al. [29] have derived the BER expressions for DPSK modulation scheme for K-distribution. They have also approximated the K-distribution expressions using Laguerre Polynomial. Xueliang Li et al. [30] have analyzed the quadrature PSK (QPSK) modulation for coherent FSO communication system and proposed a digital phase estimation technique to enhance the ABER of FSO communication system.

By combining SIM with PSK, studies indicate that the FSO system exhibits superior performance compared to other modulation schemes like QPSK, DPSK, QAM, and similar ones. This enhanced performance due to its properties such as constant amplitudes, high spectral efficiency, high receiver sensitivity, and effective background noise rejection at the expense of increased receiver complexity. Bobby Barua et al. [31] have analyzed the FSO communication system using OOK and subcarrier Binary PSK (BPSK) modulation schemes under Gamma-Gamma and negative exponential turbulence models. The BER performance of FSO communication system is found better when used with subcarrier BPSK modulation compared to OOK modulation scheme. K. Prabhu et al. [32] have derived the ABER and outage probability (OP) expressions for the BPSK-SIM and analyzed the performance for the single-input single-output (SISO) and multiple-input multiple-output (MIMO) FSO communication system and performance of the MIMO system is observed better than SISO system at the cost of complexity. In Ref. [33], M. Faridzadeh et al. have analyzed the hybrid scheme of PPM-BPSK-SIM and compared the performance of the performance with the PPM and BPSK modulation schemes over Log-Normal turbulence model and observed that the performance of the PPM-BPSK-SIM is superior to the PPM and BPSK modulation schemes.

5.3.4 QAM

The concept of transmitting a higher number of bits per symbol is achieved through the generation of symbols that combine both amplitude and phase. This modulation scheme is known as QAM. 8-QAM is one example of this, which transmits three bits per symbol using two amplitude levels and four carrier phases. Other widely used variants that transmit four, six, and eight bits per symbol are 16-QAM, 64-QAM, and 256-QAM. The constellation diagram for a 16-QAM is shown in **Figure 6**.

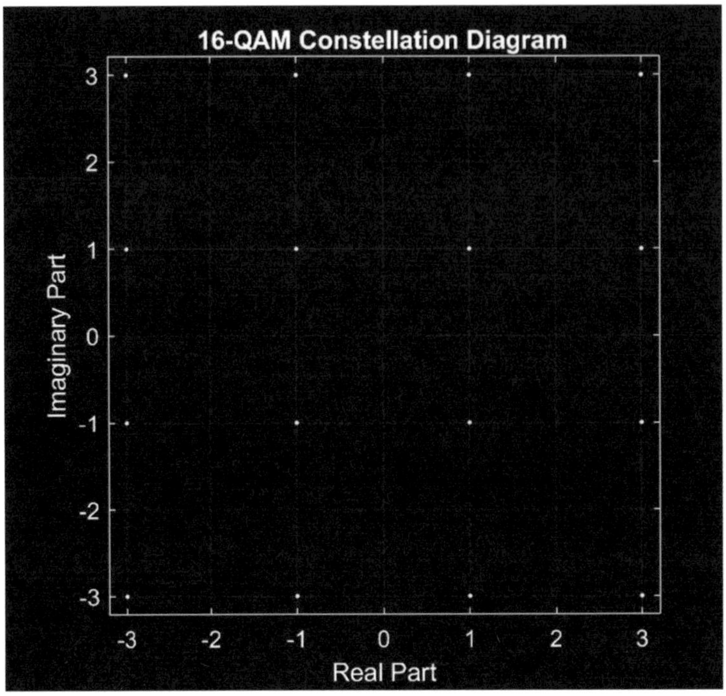

Figure 6.
16-QAM constellation diagram.

Although QAM offers remarkable spectrum efficiency, it poses challenges in demodulation, particularly when noise is present and noise is characterized by random amplitude variations. For QAM, linear power amplification is also required. Despite these considerations, QAM finds extensive usage in various technologies such as satellites, cellular phone networks, WiFi wireless LANs, and cable TV. Its utilization in these domains enables the transmission of maximum data rates within limited bandwidths.

Haitham S. Khallaf et al. [34] have analyzed the system performance for a hybrid modulation schemes, i.e. QAM and MPPM over Gamma-Gamma turbulence model under all AT regimes, and also compared the performance of the hybrid modulation QAM and MPPM with the OOK and PPM modulation schemes. They found that the system performance of the hybrid modulation QAM and MPPM is better than OOK and PPM modulation schemes. In Ref. [35], Francisco J. Escribano et al. have analyzed another hybrid modulation technique QAM and PPM and they have derived the simplified expressions for the hybrid modulation QAM and MPPM and proposed a new way to demodulate it.

5.4 OFDM

OFDM modulation is widely utilized in broadband wireless communication and is known for its resistance to frequency selective and multipath fading. By combining OFDM with FSO, one can create OFDM-based FSO, which benefits from both technologies.

The OFDM-FSO system can achieve high data rates in which a single high-rate data stream is divided into several low-rate data streams and then transmitted through narrowband subcarriers. Due to the less distortion experienced by narrowband subcarrier data streams compared to high-speed ones, there is no need for equalization. Additionally, microwave devices have a higher level of maturity compared to their optical counterparts and microwave filters exhibit superior frequency selectivity while microwave oscillators provide greater frequency stability than corresponding optical devices.

The OFDM signal for N subcarriers can be written as:

$$S_{OFDM}(t) = \sum_{n=0}^{N-1} S_n(t) \tag{19}$$

The message signal is initially modulated using one of the modulation schemes such as BPSK, QPSK, 8-PSK, 16-QAM, and 64-QAM based on the desired data rate. Then, the subcarriers are transmitted via a high-frequency carrier f_c. This mapping process is employed to convert the [19] into complex data symbols, which can be expressed as:

$$S_{OFDM}(t) = \sum_{n=0}^{N-1} X_n \left(\exp\left(j\left(w_n + 2\pi f_c t\right)\right)\right) \tag{20}$$

The equation mentioned above depicts the modulation of each symbol, X_n, onto orthogonal subcarriers. The Inverse Fast Fourier Transform (IFFT) is used to maintain orthogonality among all the subcarriers within the symbol interval. In order for the signal $X_n(t)$ to be real-valued, the IFFT input vector must have conjugate symmetry (also known as Hermitian symmetry). The first input X_0 represents the zero frequency which remains unmodulated and is typically real-valued. This approach, known as Discrete Multitone (DMT), is employed in digital subscriber line (DSL) systems [17]. Then, the optical intensity of a laser diode is modulated for transmission in free space using the valued signal, $S_{OFDM}(t)$.

Since OFDM is a multicarrier modulation technique, it strategically spaces carriers to ensure orthogonality among them, facilitating separation at the receiver and eliminating inter-symbol interference (ISI). The implementation of orthogonal carriers is achieved by using the IFFT at the transmitter. A block-type pilot arrangement is typically used for channel estimation, which is carried out by sending pilots on each sub-channel and using the estimate for a predetermined number of following symbols [36]. The input signal undergoes baseband modulation or mapping, which transforms the signal into a complex form. Subsequently, the signal that has been mapped is transformed from a serial format to a parallel one, and then the IFFT is performed to produce the OFDM symbol. In order to improve the efficiency of the system, additional cyclic prefix (CP) bits or guard bands are appended to the resulting OFDM symbol. Subsequently, the process entails the conversion from parallel to serial and the conversion from digital to analog. At the receiver, the photodiode detects the signal and initiates the reverse process, which involves converting the received OFDM symbol back into complex bit sequences through FFT conversion. Ultimately, the intricate signal is demapped in order to recover the initial sequences of bits.

In Ref. [37], authors have given a mathematical expression for ABER and OP of OFDM for radio on FSO (RoFSO) over Gamma-Gamma distribution model. In Refs. [38, 39], performance of OFDM is investigated for $\mathcal{M}$ distribution. In Ref. [39], authors compared the direct detection with the coherent detection in OFDM and the performance of the coherent detection is found better than direct detection at the cost of complexity at the receiver and authors also proposed a tunable Co-OFDM FSO system. In Ref. [40], the OFDM FSO communication for the serial decode and forward relay transmission over $\mathcal{M}$ distribution is investigated and the ABER and OP expressions are derived. Low density parity check (LDPC) coded OFDM over aggregated exponential Weibull (EW) distribution is analyzed in Ref. [41]. Authors have compared the M-*ary* QAM and M-*ary* PSK OFDM modulation schemes and the performance of QAM-OFDM is found better than PSK-OFDM. Also, they observed that the LDPC-coded OFDM gives better gain improvement compared to the uncoded OFDM system. In Ref. [42], authors have shown that the Co-OFDM helps to mitigate the effect of chromatic dispersion, polarization mode dispersion tolerance, and non-linearity in optical fiber.

5.5 FSK

FSK is utilized for transmitting digital signals with varying frequencies, where the frequency of the carrier signal is governed by the digital baseband signal. At the receiving end, the received carrier signal is transformed into a digital signal, thereby accomplishing the process of information transmission. FSK is widely employed due to its robust resistance against signal fading.

Binary FSK (BFSK) represents the simplest form of FSK. In BFSK, the frequency of the carrier wave switches between discrete binary values corresponding to the modulating signal. Consequently, the carrier frequency exhibits fluctuations based on the binary message signal. In BFSK, modulation of the carrier transpires such that a high-frequency signal is generated for a binary input of one (indicating a high level). Similarly, a low-frequency signal is produced for a binary input of zero (indicating a low level). The BFSK signal can be expressed as:

$$S(t) = A\cos\left[\left(2\pi f_c + \frac{\pi x_k}{2T}\right)t + \underbrace{\frac{\pi}{2}\sum_{n=-\infty}^{k-1} x_n - \frac{\pi}{2}(kx_n)}_{excess\ phase}\right] \tag{21}$$

where $m_1(t) = \sum_{n=-\infty}^{\infty}\overline{b}_n g(t - nT_b)$, $m_1(t) = \sum_{n=-\infty}^{\infty} b_n g(t - nT_b)$, A is the amplitude of the carrier, T_b is the period of digital symbol, b_n is sequence of numbers, and $\overline{b}_n$ is the complement of b_n.

In Ref. [43], another hybrid modulation is analyzed in which PPM and frequency shift keying (FSK) are combined and it has been shown that the ABER performance of PPM-FSK-SIM outperforms the PPM-BPSK-SIM.

5.6 MSK

MSK is a special case of continuous phase-frequency shift keying (CPFSK) characterized by a modulation index of 0.5. An index of 0.5 in MSK signifies the minimum frequency gap required to achieve coherent orthogonality between two FSK signals. The term ‘minimum shift keying’ refers to the minimum bandwidth or frequency separation required for orthogonal detection as shown in **Figure 7**. Within each symbol interval, MSK operates with one of two potential frequencies:

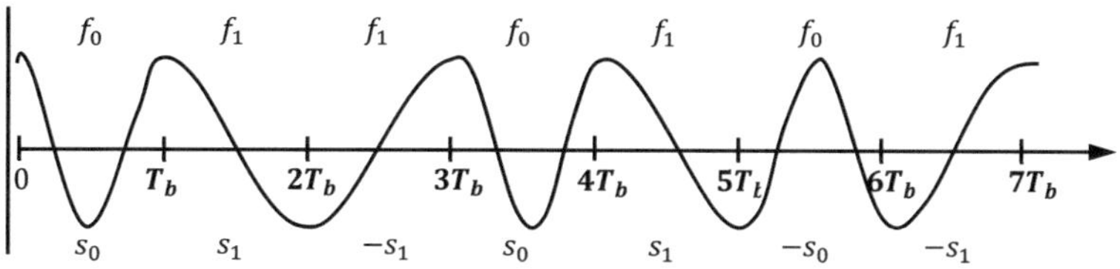

Figure 7.
MSK waveform.

$$S(t) = A\cos\left[\left(2\pi f_c + \frac{\pi x_k}{2T}\right)t + \underbrace{\frac{\pi}{2}\sum_{n=-\infty}^{k-1} x_n - \frac{\pi}{2}(kx_n)}_{excess\ phase}\right] \tag{22}$$

The principle of demodulation of MSK signal is the same as BFSK demodulation and the BER expression is given as

$$P_e = Q\left(\sqrt{\frac{2E_b}{N_0}}\right) \tag{23}$$

In Ref. [44], PPM-MSK-SIM modulation scheme is proposed. It has been compared to PPM and BPSK-SIM in the literature. Furthermore, results show that the PPM-MSK-SIM performs better for all AT regimes. Guanjun Xu also compared the PPM-MSK-SIM modulation scheme to the PPM and BPSK-SIM in Ref. [45], demonstrating that the PPM-MSK-SIM outperforms the PPM and BPSK-SIM in the Lognormal AT channel model for deep space FSO communication.

5.7 GMSK

GMSK is a modified version of MSK. A pre-modulation filter reduces a baseband pulse train's bandwidth before modulation. This filter is called the Gaussian pre-modulation filter which limits the instantaneous frequency variations by smoothing the MSK signal's phase trajectory. Consequently, a substantially smaller bandwidth FM modulated signal is obtained. The cost of this bandwidth reduction is that the pre-modulation filter smears the individual pulses in the pulse train. This temporal smearing leads to interference between adjacent pulses, known as ISI. Despite this trade-off, the advantages of GMSK in terms of power efficiency and bandwidth utilization outweigh the associated costs (**Table 1**). The BER for GMSK is:

$$P_e = Q\left(\sqrt{\frac{2\alpha E_b}{N_0}}\right) \tag{24}$$

where α is a constant related to BT_b.

When comparing the MSK and GMSK modulation schemes for short-range visible light communication (VLC), Yu Chen et al. showed that the MSK performs better than the GMSK [46]. However, hybrid modulation scheme PPM-GMSK is analyzed in Ref. [47], where it is found that the PPM-GMSK performs better in FSO systems than PPM and MSK and covers a longer link distance than the PPM and MSK modulation technique.

When $BT_b \to \infty$, it represents to MSK.

The value of BT_b	The values of α
0.25	0.68
∞	0.85

Table 1.
Values of α for different values of BT_b.

5.8 PolSK

As an alternative to APM approaches, the digital information in PolSK is encoded in the SOP of the laser source. The SOP is represented using Stokes parameters, resulting in a symbol constellation scattered across a three-dimensional (3-D) space. A digital '1' is represented by a + 45-degree linear polarized (LP) signal, while a '0' is represented by a − 45-degree LP signal.

One notable advantage of PolSK is its resilience to laser phase noise, as long as the intermediate frequency (IF) filter has sufficient bandwidth in order to prevent the conversion of phase-to-amplitude noise. Furthermore, the SOPs of the laser source remain preserved over long distances, making the successful PolSK implementation possible. Also, the PolSK signals are not affected by the excess frequency chirp produced by all-optical processing systems, in contrast to other modulation schemes such as DPSK and FSK. Also, PolSK is a form of modulation that maintains a constant envelope, making it particularly advantageous for systems that have limitations on peak power.

In Ref. [48], authors showed that the AT had less impact on the polarization states. However, in linear PolSK, the polarization coordinates must be aligned between the transmitter and receiver to work. This requirement can be avoided by using the circular PolSK (CPolSK) modulation which is proposed in Ref. [49]. CPolSK uses the two rotation states of circular polarization for binary coded signal. There is no need for polarization coordinate alignment with CPolSK as compared to linear PolSK [50, 51]. Additionally, compared to PolSK, CPolSK does not make the system any more complex. In Ref. [52], authors compared the BER performance of OOK and CPolSK, as well as the performance of CPolSK is analyzed over the Gamma-Gamma AT model, and demonstrated that CPolSK is much less sensitive to AT than OOK modulation scheme. In Ref. [53], the polarization control error was considered in their novel CPolSK model, although the advantage of coherent detection was not used. By adopting a feedback control module for polarization, the authors in Ref. [54] proposed an adaptive polarization control for coherent linear PolSK to resolve phase fluctuation and polarization offset.

5.9 Spatial modulation

Spatial Modulation (SM) has become known as a modulation method with less implementation complexity, power efficiency, and bandwidth efficiency [55, 56]. In this modulation scheme, the data rate increases logarithmically when we increase the number of transmit antennas. Here, the information is grouped into two units carrying information:

i. A group of bits determines the modulation symbol to be transmitted.

ii. Another group of bits is utilized to select the index of the transmit antenna to be activated.

Comparing SM to traditional MIMO schemes, several studies have shown that SM provides better error performance, improved spectrum efficiency (especially with a moderate number of transmit antennas), and significant reductions in transceiver complexity. This is accomplished by only allowing one transmit antenna to be active during each signaling period, with the others being silent. As a result, system complexity can be decreased and ICI can be totally avoided by using a single transmit antenna for data transfer.

5.9.1 Space Shift Keying

Space Shift Keying (SSK) represents a special case of SM where the transfer of information solely relies on antenna indices [57, 58]. This results in a simplified implementation compared to SM. Optical SSK aims to enhance spectral efficiency in both indoor and outdoor conditions.

SSK is a MIMO technique that exclusively utilizes antenna indexes for signal transmission. In the context of optical communication, optical SSK serves as the equivalent of SSK. By employing a single aperture for signal transmission during a symbol period, SSK reduces transceiver complexity and eliminates ICI issues. Moreover, while SSK requires time-limited pulse shaping, which results in bandwidth expansion, it operates with a modulation bandwidth in the range of a few GHz, making the system not limited by bandwidth constraints.

6. Conclusion

Free Space Optical (FSO) communication systems benefit from a wide range of modulation schemes, each offering unique advantages and trade-offs. IM/DD systems like OOK and SIM provide simplicity and efficiency for basic applications. Advanced schemes like PPM offer high power efficiency, while coherent detection methods such as QPSK and QAM deliver high data rates and spectral efficiency.

Emerging modulation techniques like OFDM, MSK, GMSK, PolSK, SM, and SSK further enhance the capabilities of FSO systems by improving robustness, spectral efficiency, and power efficiency. By carefully selecting and optimizing modulation schemes based on specific system requirements and channel conditions, FSO communication systems can achieve optimal performance and address the diverse needs of modern communication networks.

Thanks

The author thanks his PhD supervisors, Prof. Vijay Janyani and Prof. Ghanshyam Singh, for their motivation to write this book chapter and guidance throughout the PhD journey.

Conflict of interest

The authors declare no conflict of interest.

Author details

Bipin Kumar Saw[1*], Ghanshyam Singh[1] and Manish Tiwari[2]

1 Department of ECE, Malaviya National Institute of Technology, Jaipur, India

2 Optoelectronics and Photonics Research Lab, Department of ECE, Manipal University, Jaipur, India

*Address all correspondence to: bipin.ece@mnit.ac.in

References

[1] Andrews LC, Phillips RL. Society of Photo-Optical Instrumentation Engineers. Laser Beam Propagation through Random Media. Bellingham, Washington, USA: SPIE; 2005. 782 p

[2] Khalighi MA, Uysal M. Survey on free space optical communication: A communication theory perspective. IEEE Communications Surveys and Tutorials. 2014;**16**(4):2231-2258

[3] Majumdar Arun K. Advanced Free Space Optics (FSO) a Systems Approach. 2015. Available from: www.springer.com/series/624

[4] Zhu X, Kahn JM. Free-space optical communication through atmospheric turbulence channels. IEEE Transactions on Communications. 2002;**50**(8):1293-1300

[5] Yang L, Gao X, Alouini MS. Performance analysis of free-space optical communication systems with multiuser diversity over atmospheric turbulence channels. IEEE Photonics Journal. 2014;**6**(2):1-17

[6] Al-Habash MA, Philips RL. Mathematical model for the irradiance probability density function of a laser beam propagating through turbulent media. Optical Engineering. 2001;**40**(8):1554-1562

[7] Parry G, Puaey PN. K distributions in atmospheric propagation of laser light. Journal of the Optical Society of America. 1979;**69**(5):796-798

[8] Yilmaz F, Alouini MS. Extended Generalized-K (EGK): A New Simple and General Model for Composite Fading Channels. York, UK: Engineering, Physics; 2010.

[9] Chatzidiamantis ND, Sandalidis HG, Karagiannidis GK, Kotsopoulos SA, Matthaiou M. New results on turbulence modeling for free-space optical systems. In: ICT 2010: 2010 17th International Conference on Telecommunications. Doha, Qatar: IEEE; 2010. pp. 487-492

[10] Andrews LC, Phillips RL. I-K distribution as a universal propagation model of laser beams in atmospheric turbulence. Journal of the Optical Society of America. A. 1985;**2**(2):160-163

[11] Jurado-Navas A, Maria J, Francisco J, Puerta-Notario A. A unifying statistical model for atmospheric optical scintillation. In: Numerical Simulations of Physical and Engineering Processes. London, UK: InTech Open; 2011. pp. 181-206

[12] Cheng M, Guo Y, Li J, Zheng X, Guo L. Inverse Gaussian gamma distribution model for turbulence-induced fading in free-space optical communication. Applied Optics. 2018;**57**(12):3031-3037

[13] Peppas KP, Alexandropoulos GC, Xenos ED, Maras A. The Fischer-Snedecor F-distribution model for turbulence-induced fading in free-space optical systems. Journal of Lightwave Technology. 2020;**38**(6):1286-1295

[14] Ansari IS, Yilmaz F, Alouini MS. Performance analysis of free-space optical links over Málaga (M) turbulence channels with pointing errors. IEEE Transactions on Wireless Communications. 2016;**15**(1):91-102

[15] Badarneh OS, Mesleh R. Diversity analysis of simultaneous mmWave and free-space-optical transmission over ${\cal F}\!$F-distribution channel models. Journal of Optical Communications and Networking. 2020;**12**(11):324-334

[16] Badarneh OS, Derbas R, Almehmadi FS, El Bouanani F, Muhaidat S. Performance analysis of FSO communications over F turbulence channels with pointing errors. IEEE Communications Letters. 2021;**25**(3): 926-930

[17] Zabih G, Popoola W, Rajbhandari S. Optical Wireless Communications: System and Channel Modelling with MATLAB. Boca Raton, FL, USA: CRC Press; 2013

[18] Kedar D, Arnon S. Urban_optical_wireless_communication_networks_the_main_challenges_and_possible_solutions. IEEE Optcal Communication. 2004;**42**(5):S2-S7

[19] Zhu X, Kahn JM, Wang J. Mitigation of turbulence-induced scintillation noise in free-space optical links using temporal-domain detection techniques. IEEE Photonics Technology Letters. 2003;**15**(4):623-625

[20] Jurado-Navas A, María Garrido-Balsells J, Paris F, Castillo-Vázquez M, Puerta-Notario A. Impact of Pointing Errors on the Performance of Generalized Atmospheric Optical Channels. 2012. Available from: http://functions.wolfram.com/

[21] Huang W, Takayanagi J, Sakanaka T, Nakagawa M. Atmospheric optical communication system using subcarrier PSK modulation. In: IEEE International Conference on Communications. Geneva, Switzerland: IEEE; 1993. pp. 1597-1601

[22] Popoola WO, Ghassemlooy Z. BPSK subcarrier intensity modulated free-space optical communications in atmospheric turbulence. Journal of Lightwave Technology. 2009;**27**(8): 967-973

[23] Xu F, Khalighi MA, Bourennane S. Coded PPM and multipulse PPM and iterative detection for free-space optical links. Journal of Optical Communications and Networking. 2009; **1**(5):404-415

[24] Muhammad SS, Javornik T, Jelovčan I, Ghassemlooy Z, Leitgeb E. Optical communications comparison of hard-decision and soft-decision channel coded M-ary PPM performance over free space optical links. European Transactions on Telecommunications. 2009;**20**(8):746-757

[25] Hamkins J, Moision B. Selection of modulation and codes for deep-space optical communications. In: Free-Space Laser Communication Technologies XVI. USA: SPIE; 2004. p. 123

[26] Kiasaleh K. Performance of APD-based, PPM free-space optical communication systems in atmospheric turbulence. IEEE Transactions on Communications. 2005;**53**(9):1455-1461

[27] Xiang Y, Zengji L, Peng Y, Tao S. BER performance analysis for M ary PPM over gamma-gamma atmospheric turbulence channels. In: 2010 6th International Conference on Wireless Communications Networking and Mobile Computing (WiCOM), Chengdu, China, 2010. Chengdu, China: IEEE Antennas and Propagation Society; 2010. pp. 1-4

[28] Song X, Yang F, Cheng J, Alouini MS. Subcarrier MPSK/MDPSK modulated optical wireless communications in lognormal turbulence. In: 2015 IEEE Wireless Communications and Networking Conference (WCNC), New Orleans, LA, USA. New Orleans, LA, USA: IEEE; 2015. pp. 41-45

[29] Singh RK, Karmeshu KS. A novel approximation for K distribution:

Closed-form BER using DPSK modulation in free-space optical communication. IEEE Photonics Journal. 2017;**9**(5):1-14

[30] Li X, Geng T, Ma S, Li Y, Gao S, Wu Z. Performance improvement of coherent free-space optical communication with quadrature phase-shift keying modulation using digital phase estimation. Applied Optics. 2017; **56**(16):4695

[31] Barua B. Comparison the performance of free-space optical communication with OOK and BPSK modulation under atmospheric turbulence. International Journal of Engineering Science and Technology (IJEST). 2011;**3**(5):4391-4399. Available from: https://www.researchgate.net/publication/260245113

[32] Prabu K, Sriram Kumar D, Malekian R. BER analysis of BPSK-SIM based SISO and MIMO FSO systems in strong turbulence with pointing errors. Optik (Stuttg). 2014;**125**(21): 6413-6417

[33] Faridzadeh M, Gholami A, Ghassemlooy Z, Rajbhandari S. Hybrid pulse position modulation and binary phase shift keying subcarrier intensity modulation for free space optics in a weak and saturated turbulence channel. Journal of the Optical Society of America A. 2012;**29**(8):1680-1685

[34] Khallaf HS, Shalaby HMH, Garrido-Balsells JM, Sampei S. Performance analysis of a hybrid QAM-MPPM technique over turbulence-free and gamma-gamma free-space optical channels. Journal of Optical Communications and Networking. 2017; **9**(2):161-171

[35] Escribano FJ, Wagemakers A. Performance analysis of qam-mppm in turbulence-free fso channels: Accurate derivations and practical approximations. IEEE Systems Journal. 2021;**15**(2):1753-1763

[36] Richard VN, Ramjee P. OFDM for Wireless Multimedia Communications. London, UK: Artech House universal personal communication library; 2000

[37] Bekkali A, Ben NC, Kazaura K, Wakamori K, Matsumoto M. Transmission analysis of ofdm-based wireless services over turbulent radio-on-fso links modeled by gamma—Gamma distribution. IEEE Photonics Journal. 2010;**2**(3):510-520

[38] Duan M, Wang P, Liu X, Li Y, Chen W, Li A. Average bit error rate performance analysis of a low-density parity-check-coded orthogonal frequency-division multiplexing FSO system under Málaga distribution considering atmospheric attenuation and pointing errors. Applied Optics. 2018; **57**(19):5505

[39] Nistazakis HE, Stassinakis AN, Sandalidis HG, Tombras GS. QAM and PSK OFDM RoFSO over M-turbulence induced fading channels. IEEE Photonics Journal. 2015;**7**(1):1-11

[40] Balaji KA, Praveen BK. Performance analysis of relay assisted multihop coherant OFDM system over Malaga distribution with pointing errors. Journal of Optical Communications. 2021;**42**(3): 545-554

[41] Liu X, Wang P, Liu T, Li Y, Guo L, Tian H. ABER performance of LDPC-coded OFDM free-space optical communication system over exponentiated Weibull fading channels with pointing errors. IEEE Photonics Journal. 2017;**9**(4):1-13

[42] Wang Y, Wang D, Ma J. On the performance of coherent OFDM Systems in free-space optical communications. IEEE Photonics Journal. 2015;7(4):1-10

[43] Dubey D, Prajapati YK, Tripathi R. Error performance analysis of PPM-and FSK-based hybrid modulation scheme for FSO satellite downlink. Optical Quantum Electronics. 2020; **52**(6):1-16

[44] Liu H, Liao R, Wei Z, Hou Z, Qiao Y. BER analysis of a hybrid modulation scheme based on PPM and MSK subcarrier intensity modulation. IEEE Photonics Journal. 2015;7(4):1-10

[45] Xu G. BER and channel capacity of a deep space FSO communication system using L-PPM-MSK-SIM scheme during superior solar conjunction. Optics Express. 2019;**27**(17):24610

[46] Chen Y, Li Z, Liu P, Liu S. The BER performance comparison of MSK and GMSK schemes for short-range visible light communication. In: 2017 IEEE 9th International Conference on Communication Software and Networks (ICCSN). Beijing, China: IEEE; 2017. pp. 611-614

[47] Sahoo PK, Prajapati YK, Tripathi R. PPM- and GMSK-based hybrid modulation technique for optical wireless communication cellular backhaul channel. IET Communications. 2018;**12**(17):2158-2163

[48] Chi N, Xu L, Yu S, Jeppesen P. Generation and transmission performance of 40 Gbit/s polarisation shift keying signal. Electronics Letters. 2005;**41**(9):547-549

[49] Zhao X, Yao Y, Sun Y, Liu C. Circle polarization shift keying with direct detection for free-space optical communication. Journal of Optical Communications and Networking. 2009; **1**(4):307-312

[50] Prabu K, Kumar DS. MIMO free-space optical communication employing coherent BPOLSK modulation in atmospheric optical turbulence channel with pointing errors. Optics Communication. 2015;**343**:188-194

[51] Chao Liu CL, Yong Yao YY, Yanfu Yang YY, Yijun Yuan YY, Yufeng Zhao YZ, Benshuang Yu BY. Performance of free-space optical communication systems using circle polarization shift keying with spatial diversity receivers. Chinese Optics Letters. 2013;**11**(s2):S20101-S320104

[52] Hong Y, Yuan X, Zhang Y, Zhang Y, Wang L, Ma H. Circular polarization shift keying with homodyne coherent detection in gamma-gamma atmospheric turbulence channel. In: 2015 International Conference on Optical Instruments and Technology: Advanced Lasers and Applications. USA: SPIE; 2015. p. 96210A

[53] Wang Y, Du F, Ma J, Tan L. Employing circle polarization shift keying in free space optical communication with gamma-gamma atmospheric turbulence channel. Optics Communication. 2014;**333**:167-174

[54] Ding S, Li R, Luo Y, Dang A. Polarization coherent optical communications with adaptive polarization control over atmospheric turbulence. Journal of the Optical Society of America A. 2018;**35**(7): 1204-1211

[55] Mesleh R, Haas H, Ahn CW, Yun S. Spatial modulation—A new low complexity spectral efficiency enhancing technique. In: 2006 First International Conference on Communications and

Networking in China. USA: IEEE; 2006. pp. 1-5

[56] Mesleh R, Elgala H, Haas H. Optical spatial modulation. Journal of Optical Communications and Networking. 2011; **3**(3):234-244

[57] Jeganathan J, Ghrayeb A, Szczecinski L, Ceron A. Space shift keying modulation for MIMO channels. IEEE Transactions on Wireless Communications. 2009;**8**(7):3692-3703

[58] Abaza M, Mesleh R, Mansour A, Aggoune EHM. In: Dingel BB, Tsukamoto K, editors. The Performance of Space Shift Keying for Free-Space Optical Communications over Turbulent Channels. Proceedings of SPIE; 2015. p. 93870V

Chapter 8

Design of Optical Devices Based on a 1D Birefringent Dielectric Photonic Crystal

Noama Ouchani

Abstract

Photonic crystals (PCs) made of anisotropic dielectric materials support electromagnetic waves of coupled polarization. This wave-coupling property can be used in the design of important optical devices. We demonstrate that a one-dimensional structure composed of anisotropic dielectric materials can be used to design a transverse-electric/transverse-magnetic polarization converter. The major benefit of this converter is the combination of an identical and very high polarization conversion efficiency for both reflection and transmission waves. Furthermore, we investigate the fact that for an appropriate choice of material and geometrical properties, the birefringent photonic crystal can exhibit an absolute photonic band gap (PBG) in which light propagation is forbidden for all polarizations and for a given plane of incidence. We also explore the possibility of exploiting a single-defect anisotropic photonic crystal to fabricate an optical switch and a selective electromagnetic filter. In these optical devices, the birefringence of the photonic crystal layers increases the performance of the optical device compared with the isotropic structure.

Keywords: photonic crystal, dielectric material, anisotropic media, photonic band gap, defect modes, optical device

1. Introduction

Photonic crystals (PCs) are structures with dielectric permittivity that varies periodically on the wavelength scale in one or more directions in space [1]. This periodic modulation of the optical index leads to the formation of frequency regions (band gaps) where the propagation of electromagnetic waves is prohibited. Since the original work of Yablonovitch [2] and John [3], photonic crystals known as photonic bandgap materials have expanded considerably to cover a wide range of the electromagnetic spectrum, from microwaves to the visible. However, numerous technological difficulties limit the fabrication of three-dimensional (3D) PCs in the visible and infrared regions due to their small lattice constant, which is assumed to be comparable to wavelength [4]. This complication relating to 3D PCs led to the investigation of one-dimensional (1D) periodic structures, which can be easily produced for this range of wavelengths, using thin-film deposition techniques. Currently, 1D photonic crystals have emerged as promising materials for optical and optoelectronic components. Such components can be used to design a wide range of optical devices, including optical polarization converters [5], dielectric reflectors [6–9], and many other photonic devices, which allow all information to be processed in the form of a light wave as they can miniaturize current circuits and hence limit their cost.

The objective of this chapter is to theoretically investigate how 1D photonic crystals composed of anisotropic dielectric materials can improve the performance of optical devices compared to their counterparts made of 1D isotropic photonic crystals. The effect of dielectric anisotropy inherent in almost all natural materials, although weak in many dielectric materials, is able to perturb the polarization states of optical waves when propagating in an anisotropic photonic crystal. This means that the polarization of the emitted optical wave will be different from the polarization of the input wave and may even be orthogonal to it. In fact, the 1D birefringent photonic crystal behaves like an optical polarization converter for a given frequency value and angle of incidence [5]. This polarization conversion property has been studied in various contexts. Gao et al. [10] have proposed an ultra-wideband linear polarization converter formed of a top metasurface, an intermediate dielectric layer, and a bottom continuous metallic. Their results in the microwave frequencies revealed that the structure's metasurface can convert linearly polarized waves into their cross-polarized waves with an average reflection polarization conversion rate of 90%. Tremain et al. [11] demonstrated that a metamaterial consisting of a perforated metallic layer separated from a ground plane by a dielectric could convert the polarization of optical waves by reflection at a rate of 95%, and in Ref. [12], broadband polarization conversion has been achieved from a finite metallic grating in the microwave regime. In these works, the authors focused their attention on the structure's surface morphology to design their optical polarization converter by reflection or transmission with a high level of performance. In this chapter, we present a synthesis study of reference [5] that focuses on the design of a polarization converter based on a 1D birefringent dielectric photonic crystal. This optical device can convert wave polarization by reflection and transmission, with a high and identical polarization conversion efficiency for transverse electric (TE) and transverse magnetic (TM) inputs.

The property of birefringence can also enhance other properties of optical devices, such as reflection and selective filtering. With appropriate design, some PCs can behave as omnidirectional reflectors (ODRs) or filters. ODRs based on alternating isotropic and/or anisotropic dielectric layers, exhibiting perfect reflection at all angles of incidence and for both transverse-electric and transverse-magnetic polarizations over a wide spectral range, have been extensively studied over the last few decades. Kumar et al. [13] investigated theoretically that a 1D isotropic dielectric structure can exhibit total omnidirectional reflection (ODR) of the incident wave for a wide range of wavelengths with an appropriate choice of the values of the structure's control parameters. Chacko et al. [14] examined the effect of material dispersion on omnidirectional reflector properties in the case of a 1D isotropic dielectric PC. Their results showed that the ODR bandwidth was reduced with the inclusion of material dispersion. Omnidirectional band gaps (OBGs) can be achieved through heterostructures formed by a cascade of simple 1D photonic crystals [15] or by a cascade of ternary 1D photonic crystals composed of three types of material with different refractive indices [16]. In addition, a considerable amount of research has been carried out using 1D quasi-periodic dielectric structures to achieve omnidirectional reflectors [17, 18]. Xiao et al. [17] studied that a Fibonacci quasi-periodic PC containing metallic and dielectric layers produces an omnidirectional reflection bandgap with a width of 241 nm for all angles of incidence and both TE and TM polarizations. As for 1D periodic heterostructures, the combination of two or more quasi-periodic PCs has been shown to be an effective way of extending the bandwidth of omnidirectional reflection, as illustrated by Jiang [18].

In all the above works, isotropic dielectric materials have been used in omnidirectional reflectors. However, a few works [8, 9, 19] have focused on anisotropic dielectric materials to improve the performance of omnidirectional reflectors. Cojocaru [9, 19] proved that the omnidirectional reflection band is much wider for the birefringent structure than for the isotropic one, as well as his results revealing that two narrow ODR bands are achieved in the same spectral range from birefringent Fibonacci multilayers composed of alternating isotropic and uniaxial layers. The property of omnidirectional reflection can occur with periodic or quasi-periodic anisotropic systems. In our previous work [7], we demonstrated that a cladded-birefringent photonic crystal can exhibit an absolute photonic band gap in which light propagation is forbidden for all polarizations and for a given plane of incidence. A synthesis of this research is presented in a section of this chapter.

The engineering design of an optical device can be improved by breaking the periodicity of a perfect PC structure. This periodicity can be disrupted by adding a new layer inside the structure, changing the thickness of a layer, or removing a layer from the perfect photonic crystal. A defect cavity of a defective PC introduces defect modes localized within the photonic band gaps of the perfect structure. These modes are clearly observed as resonant peaks in the transmission/reflection spectra of the structure, allowing the wave to propagate at a certain frequency while blocking waves at all other frequency ranges. This property of defective PC has been exploited in the design of a wide range of optical devices [20–25], including optical filters [20–22], sensors [23, 24], demultiplexers [25], and so on. Since the great potential of these structures in the manufacture of photonic devices was revealed, several studies have been devoted to the selection of special materials for the defect layer. Depending on such combinations of 1D photonic structures, many types of defect layers have been considered, such as dielectric materials [6, 20, 21], metamaterials [26], superconductors [27], graphene sheets [28, 29], and so on. In this chapter, we present the design of an optical filter based on 1D dielectric photonic crystal with a birefringent defect layer. It has been shown that the properties of the defect modes depend strongly on the birefringence of the defect layer. In addition, the behavior of the anisotropic defect cavity is more complex than that observed in 1D defective PC made of isotropic dielectric material [6]. In fact, this structure can act as an optical switch: when the TE and TM polarization modes are decoupled, only one mode can pass with an intensity almost equal to 1.

This chapter is organized as follows: in Section 2, we present the theoretical model for deriving the optical properties of the proposed devices using Green's function method. Then, in the following sections, we discuss the design of the TE/TM polarization converter (Section 3), the optical reflector (Section 4), the selective electromagnetic filter, and the optical switch (Section 5). Finally, a summary of this work is given in Section 6.

2. Theoretical model

The geometrical structure of a 1D birefringent dielectric photonic crystal is illustrated schematically in **Figure 1** and consists of alternating layers of two birefringent

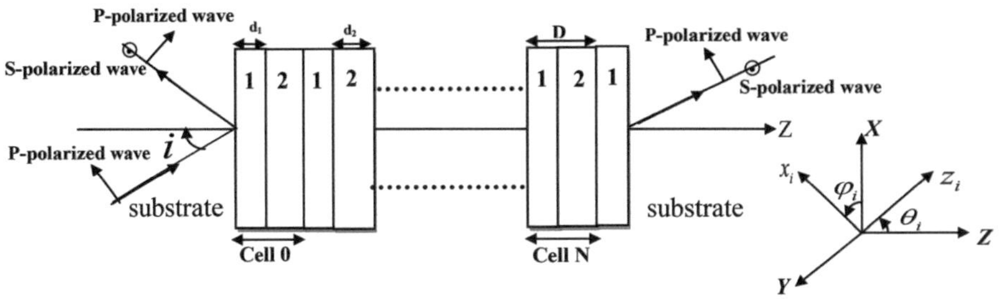

Figure 1.
Schematic representation of a finite 1D birefringent dielectric photonic crystal composed of N cells. The unit cell of the structure is formed by two birefringent materials with low (1) and high (2) refractive indices. Layers 1 and 2 are characterized by their thicknesses d_1 and d_2. $D = d_1 + d_2$ is the period of the structure. The main axes (x_i, y_i, z_i) of the layers ($i \equiv 1$ or 2) are oriented according to the azimuthal angle $\phi_{1/2}$ and the polar angle $\theta_{1/2} =$ 0° or 90° with respect to the laboratory axes (X, Y, Z). All layer interfaces are parallel to the (XY) plane of a Cartesian laboratory coordinate system, and the Z axis lies along the normal to the interfaces. The input and output substrates are isotropic media, and the incoming light propagates in the (YZ) plane with frequency ω and angle of incidence i.

materials with low (1) and high (2) refractive indices. The finite structure is embedded between two semi-infinite isotropic substrates. We assume that the materials forming the layers of the PC are homogeneous, non-magnetic, and characterized by their thicknesses $(d_{1/2})$ and their diagonal dielectric tensor $\overleftrightarrow{\epsilon}_{1/2} = \left(\epsilon_x^{1/2}, \epsilon_y^{1/2}, \epsilon_z^{1/2}\right)$, where $\epsilon_x^{1/2}$, $\epsilon_y^{1/2}$ and $\epsilon_z^{1/2}$ are the dielectric constants in the corresponding principal axes $\left(x_{1/2}, y_{1/2}, z_{1/2}\right)$ of the crystal. For a given orientation of each crystal, its dielectric tensor is described in the Cartesian laboratory coordinate system (X, Y, Z) by the Euler angles $\psi_{1/2}$, $\theta_{1/2}$, and $\phi_{1/2}$ [30, 31]. For reasons of simplicity and without loss of generality, in this study, we assume $\psi_{1/2} = 0°$, $\theta_{1/2} = 0°$ or 90° and we take the azimuthal angle $\phi_{1/2}$ as a variable. The axes of the birefringent layers are shown in **Figure 1**, and the number of the periods $(D = d_1 + d_2)$ of the structure is N.

Among the most widely used approaches for studying electromagnetic wave propagation in lamellar periodic structures are the transfer matrix and Green's function methods. In this study, we briefly recall the theory of interface response of continuous media, which aims to calculate the Greens function of a composite system containing a large number of interfaces separating different homogeneous media. Within the framework of this theory, the Green's function of a composite system can be written as [32, 33]:

$$g(DD) = G(DD) + G(DM)\left\{[G(MM)]^{-1}g(MM)[G(MM)]^{-1} - [G(MM)]^{-1}\right\}G(MD), \tag{1}$$

where D and M are, respectively, the whole space and the space of the interfaces in the lamellar system. G is a block-diagonal matrix in which each block G_i corresponds to the bulk Green's function of the subsystem i. All the matrix elements $g(DD)$ of the composite material can be obtained from the knowledge of the matrix elements $g(MM)$ in the interface space M. The $g(MM)$ is calculated from its inverse $g^{-1}(MM)$. The latter is formed out by a linear superposition of the surface matrix elements $g_{si}^{-1}(MM)$ of any independent film bounded by perfectly free interfaces with appropriate boundary conditions. The detailed calculation of the matrix elements $g_{si}^{-1}(MM)$ for a semi-infinite anisotropic medium (substrate) and an anisotropic thin film is given in previous works [7, 34]. We briefly recall these expressions, as well as those for an isotropic substrate and an isotropic thin film [32, 33]:

- The inverse of Green's function $g_{si}(M_i, M_i)$ in the space of interfaces $M_i \equiv \left[\frac{-d_i}{2}, \frac{+d_i}{2}\right]$ of an anisotropic thin film is written [7, 34]:

$$g_{si}^{-1}(M_i, M_i) = \begin{pmatrix} A_i & B_i \\ B_i & A_i \end{pmatrix} \tag{2}$$

where A_i and B_i sont (2×2) matrices, whose elements can be written in the form [7, 34]:

$$A_i = \begin{pmatrix} r_i & q_i \\ q_i & k_i \end{pmatrix} \quad et \quad B_i = \begin{pmatrix} h_i & f_i \\ f_i & e_i \end{pmatrix} \tag{3}$$

where r_i, q_i, k_i, h_i, f_i, and e_i are defined by [7, 34]:

$$\begin{aligned} r_i &= -C_i[\alpha_{i+}A_{i-}\coth\theta_{i+} - \alpha_{i-}A_{i+}\coth\theta_{i-}] \\ q_i &= C_iB_i[\alpha_{i+}\coth\theta_{i+} - \alpha_{i-}\coth\theta_{i-}] \\ k_i &= -C_iD_i[\alpha_{i+}A_{i+}\coth\theta_{i+} - \alpha_{i-}A_{i-}\coth\theta_{i-}] \\ h_i &= C_i\left[\alpha_{i+}A_{i-}(\sinh\theta_{i+})^{-1} - \alpha_{i-}A_{i+}(\sinh\theta_{i-})^{-1}\right] \\ f_i &= C_iB_i\left[\alpha_{i-}(\sinh\theta_{i-})^{-1} - \alpha_{i+}(\sinh\theta_{i+})^{-1}\right] \\ e_i &= -C_iD_i\left[\alpha_{i-}A_{i-}(\sinh\theta_{i-})^{-1} - \alpha_{i+}A_{i+}(\sinh\theta_{i+})^{-1}\right] \end{aligned} \tag{4}$$

with:
$A_{i\pm} = \left(q_Y^2 - q_0^2\epsilon_{XX}^{(i)} - \alpha_{i\pm}^2\right)$, $B_i = q_0^2\epsilon_{XY}^{(i)}$, $C_i = \left(\alpha_{i+}^2 - \alpha_{i-}^2\right)^{-1}$,
$D_i = q_0^2\epsilon_{ZZ}^{(i)}\left(q_Y^2 - q_0^2\epsilon_{ZZ}^{(i)}\right)^{-1}$, $\theta_{i\pm} = \alpha_{i\pm}d_i$ (d_i is the thickness of the layer i), $q_0 = \frac{\omega}{c}$ (q_0 is the vacuum wave vector, c is the velocity of light in vacuum, and ω is the frequency of the optical wave), and $\alpha_{i\pm}^2 = -q_Z^2$ (q_Z is the Z component of the wave vector $\vec{K} = (0, q_Y, q_Z)$ of the optical wave).

- The inverse of Green's function $g_{si}(M_i, M_i)$ in the space of interfaces of an isotropic thin film is written [32, 33]:

$$g_{si}^{-1}(M_i, M_i) = \begin{pmatrix} -\frac{F_iC_i}{S_i} & \frac{F_i}{S_i} \\ \frac{F_i}{S_i} & -\frac{F_iC_i}{S_i} \end{pmatrix} \tag{5}$$

where $C_i = cosh\,(\alpha_i d_i)$, $S_i = sinh\,(\alpha_i d_i)$ and F_i takes two different values:

$$F_i = \alpha_i \qquad for\ TE$$
$$F_i = -\frac{q_0^2 \epsilon_i}{\alpha_i} \qquad for\ TM, \tag{6}$$

in both cases: $\alpha_i^2 = k_\parallel^2 - q_0^2 \epsilon_i$

- The inverse of the Green's function $g_i(0,0)$ of the anisotropic substrate is defined by [7, 34]:

$$g_i^{-1}(0,0) = \frac{1}{(\alpha_{i+} + \alpha_{i-})} \begin{pmatrix} -(k_i^2 + \alpha_{i+}\alpha_{i-}) & B_i \\ B_i & -D_i(k^2 - \alpha_{i+}^2 - \alpha_{i-}^2 - \alpha_{i+}\alpha_{i-}) \end{pmatrix} \tag{7}$$

with $k_i^2 = q_Y^2 - q_0^2 \epsilon_{XX}^{(i)}$ and the expressions for D_i, B_i and α_{i+}, α_{i-} are given in Eq. (4).

- The inverse of the Green's function $g_i(0,0)$ of the isotropic substrate i, has the following expression [32, 33]:

$$g_i(0,0) = -F_i \tag{8}$$

These various elements of the Green's function of anisotropic and isotropic media (infinite, semi-infinite, and thin film) enable us to obtain different physical properties of the multilayer structures such as dispersion relations, reflection and transmission coefficients, delay time, and total and local densities of states of the optical modes.

2.1 Dispersion relation of an infinite PC

All the information about the propagation of electromagnetic waves in a given medium is contained in its band structure or dispersion diagram, deduced from the dispersion relation. In the following, we present the dispersion relation, characterizing the propagation of electromagnetic waves in an infinite birefringent dielectric PC, which is deduced from the interface response theory. Within the total interface space of the structure, the inverse of the matrix giving all the interface elements of the Greens function $g(MM)$ is an infinite tridiagonal matrix formed by the linear juxtaposition of the elements $g_{si}^{-1}(M_i, M_i)$ (Eq. (2)). We apply the Fourier transformation in the elementary cell of the tridiagonal matrix to obtain the following dispersion relation for the (YZ) incidence plane [7, 34]:

$$\alpha \cos^2(k_B D) + 2\beta \cos(k_B D) + \delta = 0 \tag{9}$$

with

$$\alpha = (e_2h_2 - f_2^2)(e_1h_1 - f_1^2)$$
$$\beta = 2f_1f_2[f_1^2 + f_2^2 - (r_1+r_2)(k_1+k_2)] + (f_2e_1 + f_1e_2)[2(q_1+q_2)(r_1+r_2) - h_1f_1 - h_2f_2]$$
$$+(f_2h_1 + f_1h_2)[2(q_1+q_2)(k_1+k_2) - e_1f_1 - e_2f_2] + h_1h_2\left[e_1^2 + e_2^2 - (k_1+k_2)^2\right]$$
$$+e_1e_2\left[h_1^2 + h_2^2 - (r_1+r_2)^2\right] - (q_1+q_2)^2[h_2e_1 + h_1e_2 + 2f_1f_2]$$
$$\delta = \left[(r_1+r_2)(k_1+k_2) - (q_1+q_2)^2\right]^2 - [(r_1+r_2)e_1 - (q_1+q_2)f_1]^2 - [(r_1+r_2)e_2$$
$$-(q_1+q_2)f_2]^2 - [(q_1+q_2)f_1 - (k_1+k_2)h_1]^2 - [(q_1+q_2)f_2 - (k_1+k_2)h_2]^2 + [e_1h_1 - f_1^2]^2$$
$$+[e_1h_2 + e_2h_1]^2 + [e_2h_2 - f_2^2]^2 - 2(r_1+r_2)(k_1+k_2)(f_1^2 + f_2^2) + 2(q_1+q_2)(r_1+r_2).$$
$$(e_1f_1 + e_2f_2) - 2(q_1+q_2)^2(h_1e_1 + h_2e_2) + 2(q_1+q_2)(k_1+k_2)(h_1f_1 + h_2f_2)$$
$$+4f_1f_2[f_1f_2 - h_2e_1 - h_1e_2] \tag{10}$$

where $r_i, q_i, k_i, h_i, f_i,$ and e_i are given in Eq. (4), k_B is the Bloch wave vector of the periodic system along the axis of the PC, and D is the period of the PC.

It is worth noting that the dispersion relation is of second order when the structure is formed of alternating biaxial layers with principal axes oriented in arbitrary directions. This means that there is a coupling between the two polarization modes, TE and TM. In particular, in cases where the principal axes of the biaxial layers are parallel or perpendicular to the axes of the laboratory's Cartesian coordinate system, the electromagnetic wave is split into two independent modes (TM and TE), and the dispersion relation becomes:

$$\cos(k_BD) = \cosh(\alpha_1d_1)\cosh(\alpha_2d_2) + \frac{1}{2}\left(\frac{F_1}{F_2} + \frac{F_2}{F_1}\right)\sinh(\alpha_1d_1)\sinh(\alpha_2d_2) \tag{11}$$

with,

$$\begin{aligned} \alpha_i^2 &= q_Y^2 - q_0^2\epsilon_{XX}^i && \textit{and} \quad F_i = \alpha_i && \textit{for le mode TE} \\ \alpha_i^2 &= \frac{\epsilon_{YY}^i}{\epsilon_{ZZ}^i}\left(q_Y^2 - q_0^2\epsilon_{ZZ}^i\right) && \textit{and} \quad F_i = -\frac{q_0^2\epsilon_{YY}^i}{\alpha_i} && \textit{for le mode TM} \end{aligned} \tag{12}$$

This dispersion relation remains valid when the photonic crystal is formed by isotropic media, assuming $\epsilon_{XX} = \epsilon_{YY} = \epsilon_{ZZ} = \epsilon$ in Eq. (12). Note that an in-depth calculation of the dispersion relation of the anisotropic photonic crystal calculated within the framework of the Green's function method and the transfer matrix method is given in Ref.s [7, 34, 35].

2.2 Reflection and transmission coefficients

In the context of the interface response theory, the reflected and transmitted waves $u(D)$, derived from a uniform plane wave $U(D)$ incident on a plane interface between two different media, are expressed as follows [32, 33]:

$$u(D) = U(D) + G(DM)\left\{[G(MM)]^{-1}g(MM)[G(MM)]^{-1} - [G(MM)]^{-1}\right\}U(M). \tag{13}$$

Let us mention that each incident wave propagating inside the anisotropic medium generates two transmitted waves and two reflected waves with different polarizations, namely, transverse electric TE (E ⊥ plane of incidence) or transverse magnetic TM (E || plane of incidence). Let us call E_{iS} and E_{iP} the amplitudes of the $S(TE)$ and $P(TM)$ components of the incident field, respectively. Then, the amplitudes of the reflected and transmitted fields can be written as:

$$\overleftarrow{E}_{RS}(z) = r_{SS}\overleftarrow{E}_{iS}(z) + r_{SP}\overleftarrow{E}_{iP}(z), \tag{14}$$

$$\overleftarrow{E}_{TS}(z) = t_{SS}\overleftarrow{E}_{iS}(z) + t_{SP}\overleftarrow{E}_{iP}(z), \tag{15}$$

$$\overleftarrow{E}_{RP}(z) = r_{PS}\overleftarrow{E}_{iS}(z) + r_{PP}\overleftarrow{E}_{iP}(z), \tag{16}$$

$$\overleftarrow{E}_{TP}(z) = t_{PS}\overleftarrow{E}_{iS}(z) + t_{PP}\overleftarrow{E}_{iP}(z). \tag{17}$$

The expressions of r_{ij} and t_{ij} in these equations (i, j = S or P) for an anisotropic structure are given in the previous works [7, 34].

Therefore, the total reflection coefficients for the P and S modes are given, respectively, by [7, 34]:

$$R_P = R_{PP} + R_{SP} = \left|r_{PP}\right|^2 + \left|r_{SP}\right|^2, \tag{18}$$

$$R_S = R_{SS} + R_{PS} = \left|r_{SS}\right|^2 + \left|r_{PS}\right|^2, \tag{19}$$

and the total transmission coefficients for the P and S modes are given, respectively, by [7, 34]:

$$T_P = T_{PP} + T_{SP} = \left|t_{PP}\right|^2 + \left|t_{SP}\right|^2, \tag{20}$$

$$T_S = T_{SS} + T_{PS} = \left|t_{SS}\right|^2 + \left|t_{PS}\right|^2. \tag{21}$$

In the case where the principal axes of the anisotropic layers constituting the PC are parallel or perpendicular to the fixed axes, the polarization waves S and P propagate independently with two different propagation speeds. Then the terms R_{SP}, R_{PS}, T_{SP}, and T_{PS} cancel out, and the above relations become $R_P = R_{PP} = \left|r_{PP}\right|^2$ and $T_P = T_{PP} = \left|t_{PP}\right|^2$ for an incident wave of polarization P. Moreover, $R_S = R_{SS} = \left|r_{SS}\right|^2$ and $T_S = T_{SS} = \left|t_{SS}\right|^2$ for an incident wave of polarization S.

3. Polarization converter design

The transverse-electric/transverse-magnetic polarization converter consists of a 1D birefringent dielectric photonic crystal, embedded between two semi-infinite isotropic substrates (air). The birefringent biaxial layers 1 and 2 (**Figure 1**) consist of sodium nitrite($NaNO_2$) and antimony sulfur iodide $SbSI$ with principal optical indices [36] $n_{x_1} = 1.344, n_{y_1} = 1.411, n_{z_1} = 1.651$ and $n_{x_2} = 2.7, n_{y_2} = 3.2, n_{z_2} = 3.8$, respectively. For the sake of simplicity, we assume that the polar angles of the two layers $\theta_{1/2}$ and the azimuthal angle ϕ_1 of the weakly anisotropic layer ($NaNO_2$) are zero, while the azimuthal angle ϕ_2 of the strongly anisotropic layer ($SbSI$) is taken as a variable. The thicknesses of these layers are $d_1 = 0.3\mu m$ and $d_2 = 0.7\mu m$ $D = d_1 + d_2$ is the period of the structure (**Figure 1**). The polarization conversion property is satisfied by the anisotropy of the materials. After multiple reflections and transmissions in the anisotropic layers constituting the 1D birefringent PC, the incident plane wave produces two reflected and two transmitted waves with *TE* (or *S*) and *TM* (or *P*) polarizations. Indeed, when the main axes of the anisotropic layers are parallel or perpendicular to the fixed axes, the transmitted and reflected waves of polarization *S* and *P* remain decoupled and preserve the polarization state of the incoming wave. On the other hand, for other orientations of birefringent layers, the transmitted and reflected waves of polarization *S* and *P* are coupled and present a large mixing between *S* and *P* modes [5]. **Figure 2** shows the reflection coefficients R_{PP} and R_{SP} as a function of the reduced frequency Ω (where $\Omega = \frac{\omega D}{c}$) for different values of the azimuthal angles ϕ_2 when the incoming *P*-polarized light is launched normally on the interface separating the substrate and the $(N+1)NaNO_2/(N)SbSI$ structure. It is shown that in **Figure 2(a,c)**, the *S*- and *P*-polarized reflected waves are almost decoupled, since the reflection coefficients R_{PP} are higher than R_{SP}. However, in **Figure 2(b)**, the reflected optical wave is predominantly *S*-polarized, with the highest reflection coefficient R_{SP}. In particular, the polarization conversion reaches its maximum with a rate of 83.20% at $\Omega = 3.83$ for the mode labeled *A* on the dashed curve (**Figure 2(b)**).

In order to improve the polarization conversion rate and eventually reach the maximum limit of 100%, the study of the reflection through a superlattice (SL) containing several periods for two cases of the structure: $(N+1)SbSI/(N)NaNO_2$ and $(N+1)NaNO_2/(N)SbSI$, has been developed. **Figure 3** depicts the converted

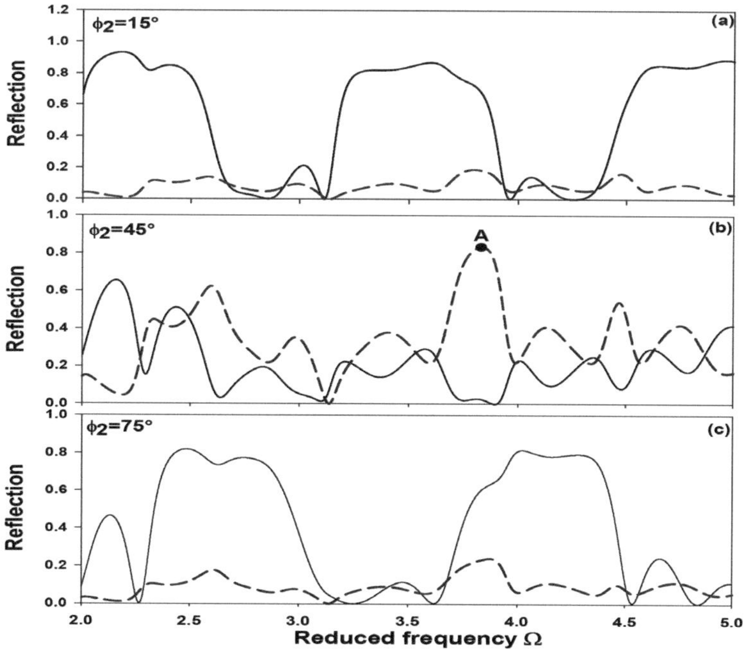

Figure 2.
Evolution of the reflectances R_{PP} (solid curves) and R_{SP} (dashed curves) versus the reduced frequency Ω for several values of the azimuth angle of the SbSI layers, namely (a) $\phi_2 = 15°$, (b) $\phi^2 = 45°$, and (c) $\phi_2 = 75°$. The structure consists of five layers of $NaNO_2$ and four layers of SbSI. The incoming P-polarized light is launched normally to the SL.

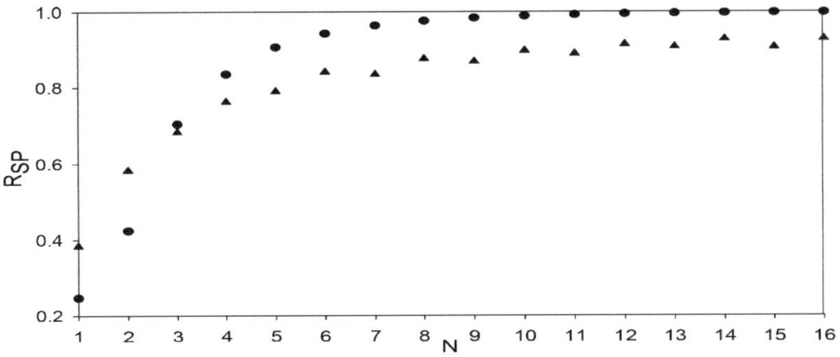

Figure 3.
Evolution of the converted reflection coefficients R_{SP} as a function of the number of cells N of the structures: $(N+1)SbSI/(N)NaNO_2$ (triangular symbols) and $(N+1)NaNO_2/(N)SbSI$ (circular symbols). The input light has a TM polarization.

reflection coefficient R_{SP} as a function of the number of periods N in the SL. The triangular or circular symbols are the maximum of R_{SP} in the frequency range $0 \leq \Omega \leq 10$, in the azimuthal angle range $0° \leq \phi_2 \leq 90°$, and for the incidence angle domain $0° \leq \theta_i < 90°$. For both types of SL, the converted reflection coefficient R_{SP} increases significantly with the number N of periods, reaching values as high as 99%. It is found that the highest conversion rates (circular symbols) are obtained for the $(N+1)NaNO_2/(N)SbSI$ structure when $N \geq 3$, reaching polarization conversion higher than 99% for $N \geq 11$. A more in-depth analysis shows that the $(12)NaNO_2/(11)SbSI$ structure, i.e., $N = 11$, exhibits a very high polarization conversion ratio of 99.22% in the frequency range $3.5 < \Omega < 7.5$ and for an azimuthal angle $\phi_2 = 48°$ and an angle of incidence $i = 10°$. Assuming $\Omega = 3.84$, **Figure 4(a,b)** illustrates the evolution of reflection coefficients R_{PP} and R_{SP} as a function of azimuthal angle ϕ_2 (**Figure 4(a)**) and incidence angle i (**Figure 4(b)**). These curves reveal the significant effect of azimuthal angle ϕ_2 and incidence angle i in achieving a high polarization ratio and, in particular, the possibility of a complete polarization conversion by reflection.

The reflection polarization converter consisting of the $(12)NaNO_2/(11)SbSI$ photonic crystal also enables polarization conversion via the transmission process. Careful analysis of the converted transmission coefficient T_{SP}, as shown in **Figure 3**, reveals

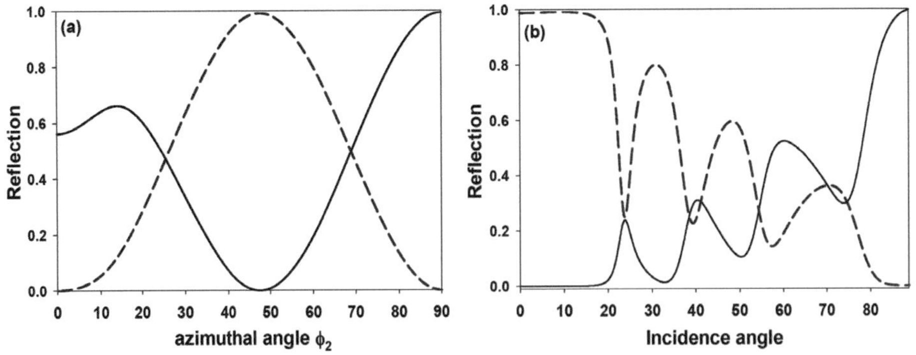

Figure 4.
Evolution of the reflectances R_{PP} (solid curves) and R_{SP} (dashed curves) as a function of azimuthal angle ϕ_2 at $(i = 10°, \Omega = 3.84)$ (a) and incidence angle i at $(\phi_2 = 48°, \Omega = 3.84)$ (b) for the SL $(12)NaNO_2/(11)SbSI$.

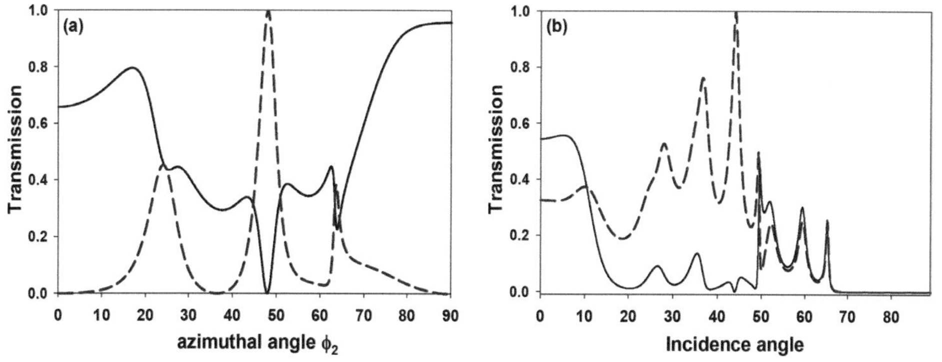

Figure 5.
Variation of the transmission coefficients T_{PP} (solid curves) and T_{SP} (dashed curves) as a function of azimuthal angle ϕ_2 at $(i = 44°, \Omega = 7.67)$ (a) and incidence angle i at $(\phi_2 = 48°, \Omega = 7.67)$ (b) for the SL $(12)NaNO_2/(11)SbSI$.

that the highest conversion ratio (99.86%) is achieved when $\phi_2 = 48°$, $\Omega = 7.67$, and $i = 44°$. Taking $\Omega = 7.67$, **Figure 5(a,b)** shows the progression of transmission coefficients T_{PP} and T_{SP} as a function of azimuthal angle ϕ_2 (**Figure 5(a)**) and incidence angle i (**Figure 5(b)**). These curves highlight the powerful effect of azimuthal angle ϕ_2 and incidence angle i in achieving a high transmission polarization ratio and, especially, in enabling full transmission polarization conversion. It is interesting to note that for any symmetrical structure such as the proposed polarization converter, the converted reflection (transmission) coefficients R_{PS} (T_{PS}) and R_{SP} (T_{SP}) behave similarly when the incoming light is of polarization S and P, respectively. This structure enables the polarization to be converted efficiently and identically for $TM \rightarrow TE$ and $TE \rightarrow TM$ by reflection and transmission [5].

In the present study, the polarization conversion process occurs as a result of the material's anisotropy, but can also be achieved by means of arbitrary surface profiles which force the input mode to decompose into two perpendicularly polarized components. The polarization converter proposed by Gao et al. [10] shows that the structure's metasurface can convert the polarized incident wave into its orthogonally polarized wave with an average reflected conversion rate of 90%. The effect of the structure's surface morphology on the polarization conversion process of the incident wave is significantly greater, as improved in Refs. [11, 12]. However, the manufacture of an optical polarization converter based on the surface morphology of a structure seems more difficult than that based on a 1D anisotropic photonic crystal. In addition, the main advantage of the TE/TM polarization converter is the possibility of combining identical and highly efficient polarization conversion for both reflection and transmission waves.

4. Optical reflector design

The optical reflector consists of a cladded finite superlattice embedded between two substrates made of an isotropic medium with a high refractive index. The cladded finite SL is formed of a 1D birefringent photonic crystal $(N)SbSI/(N)NaNO_2$ clad on one side by a biaxial layer $NaNO_2$. Each birefringent biaxial layer is characterized by its orientation defined, by the azimuthal angle ϕ_i and the polar angle θ_i with respect to the Cartesian laboratory coordinate system (X, Y, Z), and by its thicknesses $\frac{d_i}{D}$, where $D = d_1 + d_2$ is the period of the structure (**Figure 1**). The geometrical properties $\left(\phi_i, \theta_i, \frac{d_i}{D}\right)$ of the layers making up the cladded finite SL are $\left(\phi_0 = 0°, \theta_0 = 90°, \frac{d_0}{D} = 5.4\right)$ for the cladding biaxial layer $NaNO_2$ and $\left(\phi_1 = 65°, \theta_1 = 0°, \frac{d_1}{D} = 0.75\right)$ and $\left(\phi_2 = 0°, \theta_2 = 90°, \frac{d_2}{D} = 0.25\right)$ for the biaxial layers of the perfect structure $NaNO_2$ and *SbSI*, respectively. As improved in the previous work [7], the choice of these parameters reflects the conditions under which an absolute photonic bandgap is formed when light illuminates the boundary of the semi-infinite superlattice $SbSI/NaNO_2$ from the anisotropic substrate $NaNO_2$. This semi-infinite structure exhibits an absolute PBG in which optical wave propagation is prohibited for all polarizations and for every value of the wavevector $k\|$ (the area between the two horizontal lines in **Figure 6(a)**). The projected band diagram of the structure (**Figure 6**) has been derived from the analytical dispersion equation for an infinite anisotropic SL (Eq. (9)). Note that both infinite and semi-infinite PCs have the same band structure, the only difference being the presence of surface modes for a semi-infinite structure. The white and gray regions of the projected band diagram of the structure (**Figure 6**) represent the forbidden and permitted bands, respectively.

It is clear that the polar angle of the substrate $NaNO_2$ has a significant effect on the presence of an absolute PBG for both *TE* and *TM* polarizations (**Figure 6(a)**). The structure exhibits an absolute PBG when the polar angle of the substrate $NaNO_2$ differs from the polar angle of the layers $NaNO_2$ constituting the perfect PC ($SbSI/NaNO_2$). However, at the same polar angle of the substrate and the layers $NaNO_2$ of the PC, the azimuthal angle does not allow the formation of an absolute PBG for both *TE* and *TM* polarizations (**Figure 6(b)**). Thus, the weakly anisotropic layer ($NaNO_2$) is likely to act as a buffer layer at the entrance to the 1D birefringent photonic crystal, behaving as a barrier to the propagation of light generated in any substrate. In previous research [6, 15, 16], two alternatives have been proposed to overcome the limitation of substrate choice. The first involves associating the superlattice with a cladding layer of low refractive index compared with the refractive

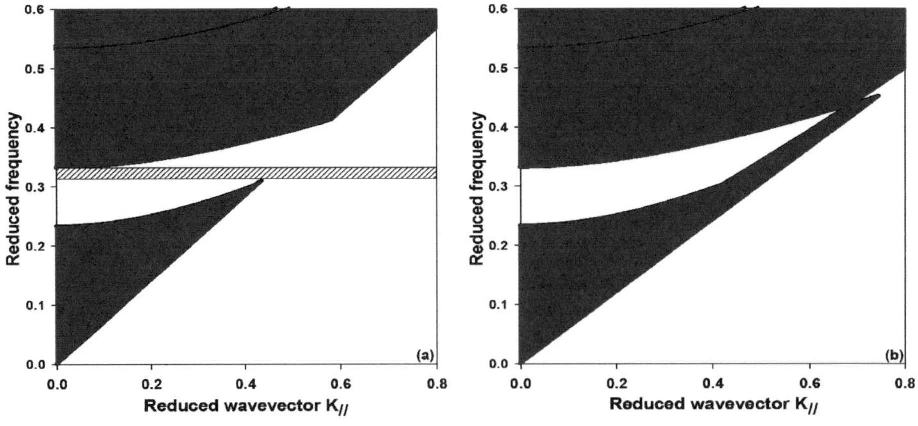

Figure 6.
Projected photonic band structure of semi-infinite anisotropic SL ($NaNO_2/SbSI$). The reduced frequency $\Omega = \frac{\omega D}{2\pi c}$ is plotted as a function of the reduced wave vector $K_\| = \frac{k_\| D}{2\pi}$. The gray and white regions correspond to the passbands and gaps of the SL, respectively. The orientation (ϕ_i, θ_i) of the substrate $NaNO_2$ is assumed to be $(0°, 90°)$ and $(0°, 0°)$ for curves (a) and (b), respectively.

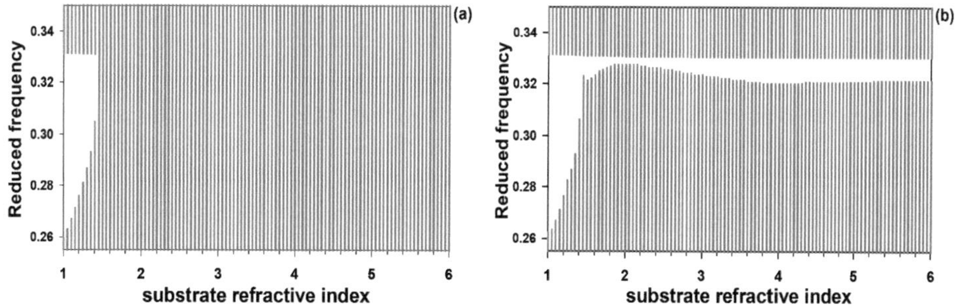

Figure 7.
Evolution of the reduced frequency Ω of the absolute PBG as a function of the refractive index of the substrate. The structural parameters of the birefringent reflector are mentioned in the text.

indices of the superlattice components [6]. The second is to combine two or more superlattices in such a way that the superposition of their band structures presents an absolute photonic bandgap [15, 16].

The effect of the buffer layer in achieving an absolute PBG, whatever the refractive index of the substrate (incident medium), is examined in **Figure** 7. This buffer layer is made of the same material ($NaNO_2$) as one of the layers, making up the perfect superlattice, but with a different orientation and thickness, as mentioned above. It is clear from **Figure 7(a)** that an absolute PBG cannot be achieved if the substrate is made of a material with a high refractive index compared to the refractive indices of the superlattice constituents. However, the inclusion of the buffer layer at the entrance to the finite superlattice (**Figure 7(b)**) ensures that the cladded finite superlattice displays an absolute photonic bandgap, regardless of the material used for the substrate. Let us note that the reflection gap in **Figure** 7 is defined as a frequency range in which the transmission intensity falls below a threshold at 10^{-3} whatever the incidence angle and the polarization of the wave. In addition, the effect of other structural parameters, namely, the orientation of the layers $NaNO_2$, the number N of unit cells in the SL, and the thickness of the buffer layer, on the width of the absolute PBG is examined in Ref. [7].

As a result, the birefringence of the layers improves the performance of the optical mirror as compared to its isotropic homolog. In fact, the birefringent reflector requires an appropriate choice of the orientations of the two materials forming the cladded finite structure in order to obtain an absolute photonic bandgap, whatever the substrate. In contrast, the isotropic reflector needs a cladding layer made of a material with a low optical index to act like a barrier to light propagation generated in the substrate, whereas this condition presents certain limitations related to the properties of the material.

5. Optical filter and switch design

In this section, we theoretically investigate the possibility of exploiting a 1D defective photonic crystal in the fabrication of optical filters and switches. The structure consists of a finite symmetrical PC (*SbSI/air*) embedded between two substrates made of an isotropic medium, with an anisotropic defect layer *SbSI* inserted in the middle of the system. The defect layer is made from the same material as the one constituting the perfect SL, but with a different orientation and thickness. In Ref.s [20, 21], we have thoroughly demonstrated the different optical transmission properties of an anisotropic cavity in a 1D isotropic PC [21] and in a 1D anisotropic PC composed of two alternating biaxial layers [20]. For reasons of simplicity, the perfect 1D anisotropic structure consists of four layers of air (material i = 1) and five layers of SbSI (material i = 2), enclosed between two isotropic media (air). The periodicity of the structure is broken by the insertion a defect layer (*SbSI*) of thickness d_0 and orientation ϕ_0 at the center of the perfect structure. We assume that the thicknesses d_1 and d_2 of the layers are chosen in accordance with the quarter wave stack condition, i.e., $n_1 d_1 = n_{z_2} d_2$, where $n_1 = 1$ is the refractive index of air and n_{z_2} is one of the principal optical indices of the anisotropic layer $n_{x_2} = 2.7, n_{y_2} = 3.2, n_{z_2} = 3.8$ [36], and the thickness of defect layer is $d_0 = 0.5D$ for which the localized modes are clearly described in PBGs, where D is the period of the SL. The orientation of the biaxial layers, making up the perfect SL, is characterized only by the azimuthal angle ϕ_2. In order to illustrate the filtering and switching behavior of localized modes in the photonic bandgaps of the structure, we assume that the incoming *S*-polarized light is launched normally on the interface separating the substrate and the structure. After repeated reflections and transmissions in the structure's anisotropic media, the incident wave produces two reflected (R_{SS} and R_{PS}) and two transmitted waves (T_{SS} and T_{PS}) with *S* and *P* polarizations, respectively. **Figure 8** displays the transmission coefficients T_{SS} and T_{PS} as a function of the reduced frequency Ω (where $\Omega = \frac{\omega D}{2\pi c}$) for a perfect PC (blue dotted lines) and a defective PC (red dashed lines and black solid lines for T_{SS} and T_{PS}, respectively). In this illustration, we suppose that the principal axes of the material (*SbSI*) constituting the perfect SL are chosen parallel to the axes of the Cartesian laboratory coordinate system (X, Y, Z), i.e., $\phi_2 = 0°$, and we take the orientation of the defect layer ϕ_0 as variable. A significant effect of the defect layer's birefringence on the intensity and number of localized modes is shown to occur

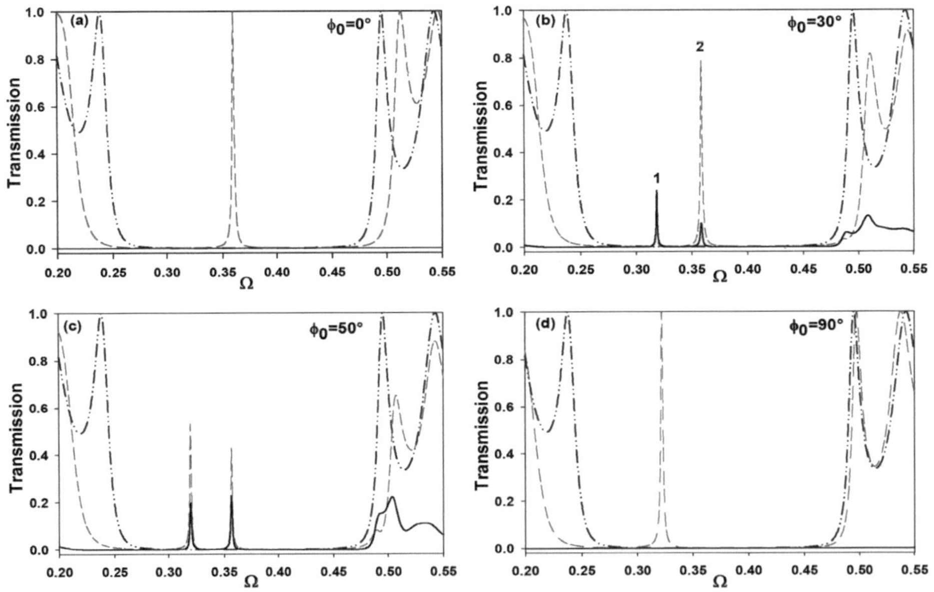

Figure 8.
Evolution of the transmission coefficients T_{SS} (red dashed lines) and T_{PS} (black solid lines) as a function of the reduced frequency Ω for $\phi_o = 0°$ (a), $\phi_o = 30°$ (b) $\phi_o = 50°$ (c), and $\phi_o = 90°$ (d). The blue dotted lines correspond to the perfect SL, and the incident S-polarized wave is launched normally to the structure.

within the lowest photonic bandgap of the transmission spectrum (**Figure 8**). When the principal axes of the defect cavity are parallel (**Figure 8(a)**) or perpendicular (**Figure 8(d)**) to the axes of the laboratory coordinate system (X, Y, Z), one mode of intensity equal to unity appears inside the PBG, with a polarization identical to that of the incident light ($T_{PS} = 0$). The situation is then similar to that of an isotropic system. For other orientations of the principal axes of defect layer ($\phi_0 = 30°$; $\phi_0 = 50°$), the S-polarized incident wave produces two transmitted waves containing both S- and P-polarized waves, with transmission coefficients T_{SS} and T_{PS}, respectively. These two propagating waves are coupled, generating two defect modes of different intensities at two different frequencies in the bandgap of the transmission spectrum of each polarization of the transmitted optical wave (**Figure 8(b,c)**).

For an in-depth analysis of the behavior of these transmitted waves, we have plotted in **Figure 9** the dispersion curves of the localized modes in the lowest PBG as a function of the orientation of the layers constituting the perfect structure ϕ_2 when the orientation of the defect layer is $\phi_0 = 0°$ (**Figure 9(a)**) and of defect layer orientations ϕ_0 when the orientations of the layers making up the perfect SL is $\phi_2 = 0°$ (**Figure 9(b)**). The branches in this figure represent the maximum transmission coefficients T_S ($T_S = T_{SS} + T_{PS}$) when the incoming light has a S polarization, while the gray regions represent the bulk bands of the infinite perfect PC. Note that the continuous bands in the infinite and semi-infinite PCs decompose into the discrete modes in the finite PC. It is clear that the behaviors of the defect modes (**Figure 9(a)**) are completely different from their behaviors when the defect layer has an arbitrary orientation (**Figure 9(b)**). A single mode appears in the PBGs with the same polarization as the incident wave, as a result of the isotropy ($\phi_0 = 0°$) of the defect layer.

The change in orientation of the birefringent defect layer produces two localized modes in the PBG, since each S- or P-polarized incident wave generates two reflected waves and two transmitted waves containing both S- and P-polarized waves. Moreover, the intensities of the localized modes are very sensitive to the birefringence of the defect layer, as illustrated in **Figure 10(a,b)**, where we have plotted the transmitted intensities T_{SS} (**Figure 10(a)**) and T_{PS} (**Figure 10(b)**) of the modes labeled 1 and 2 in **Figure 8(b)** versus the azimuthal angle ϕ_0. The finding in **Figure 10(a)** shows that by varying ϕ_0, the transmission peaks of the defect modes exhibit different behaviors within the PBGs

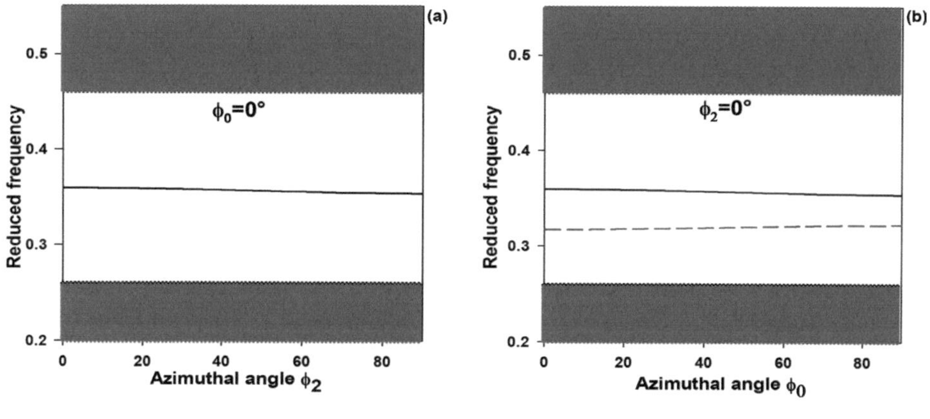

Figure 9.
Variation of the reduced frequency Ω of the defect modes as a function of the azimuthal angles ϕ_2 (a) and ϕ_0 (b). The gray regions correspond to the bulk bands of the infinite PC. The incident S-polarized wave is launched normally to the structure.

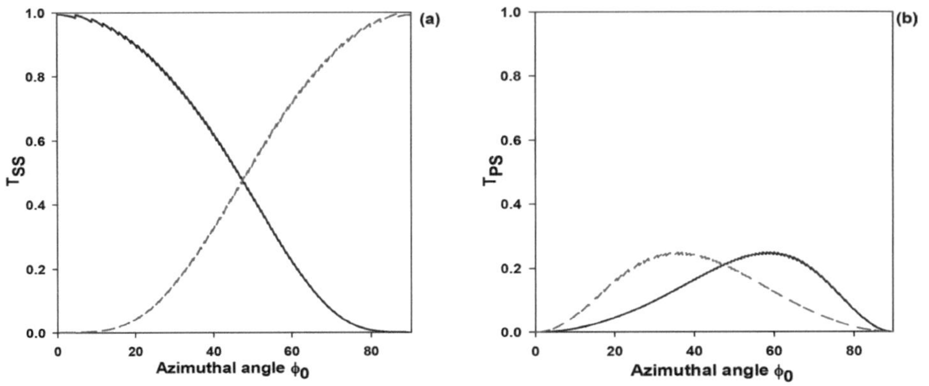

Figure 10.
*Dependence of the peak heights T_{SS} (a) and T_{PS} (b) of the defect modes labeled 1 (red dashed lines) and 2 (blue solid lines) in **Figure 8(b)**, as a function of the azimuthal angle of the defect layer ϕ_0. The incoming S-polarized light is launched normally to the SL.*

of the structure. Indeed, by rotating the azimuthal angle ϕ_0 from 0° to 90°, the transmission peaks of the localized mode 1 (red dashed line in **Figure 10(a)**) increase from 0 to its maximum value with an amplitude close to unity, while the transmission peaks of the defect mode 2 (blue solid line in **Figure 10(a)**) decrease from the unity to 0. Furthermore, the transmitted P-polarized modes (**Figure 10(b)**) display a similar behavior. Their intensities increase progressively as the azimuthal angle of the defect layer is rotated, reaching their maximum value in the region where the coupling of S and P waves is very strong and then decreasing until they disappear at $\phi_0 = 90°$. These results also remain valid for P-polarized incident waves and are in good agreement with those presented in Ref. [20] when the defective photonic crystal is composed only of anisotropic media, as well as with those presented in Ref. [21], when the defective PC is formed by a combination of a finite isotropic SL and an anisotropic defect layer. These structures offer a wide range of options for creating and controlling the number of localized modes and their frequencies and intensities. They may be good candidates for optical switches and selective electromagnetic filters.

6. Conclusion

In summary, we have theoretically investigated the possibility of exploiting one-dimensional birefringent dielectric photonic crystals in the design of optical devices. The birefringence of the structures offers a wide range of options, either for converting the polarization of incident optical waves into their orthogonal polarizations or for controlling the flow of light by creating an absolute photonic bandgap, in which the propagation of optical waves is forbidden in all directions of propagation and polarization. In addition, we have demonstrated that the birefringence of defect layers in 1D dielectric photonic crystals significantly affects the filtering and switching behaviors of localized modes in the electromagnetic bandgaps of the structures. The results show that by adjusting the birefringence properties of the structures, the performance of optical devices can be improved over their isotropic counterparts. It is worth noting that the birefringent photonic crystal may offer effective possibilities for controlling the delay time of light propagation from the subluminal to the apparently superluminal regime and vice versa, although this phenomenon has not been explored here, but was in earlier work [37].

Author details

Noama Ouchani[1,2]

1 Centre Régional des Métiers de l'Education et de la Formation, Fès, Morocco

2 LPMR, Département de Physique, Faculté des Sciences, Université Mohamed I, Oujda, Morocco

*Address all correspondence to: noama03@yahoo.fr

References

[1] Joannopoulos JD, Meade RD, Winn JN. Photonic Crystals: Molding the Flow of Light. Princeton, NJ: Princeton University Press; 1995. ISBN 10: 0691037442

[2] Yablonovitch E. Inhibited spontaneous emission on solid-state physics and electronics. Physical Review Letters. 1987;**58**(20):2059-2062. DOI: 10.1103/PhysRevLett.58.2059

[3] John S. Strong localization of photons in certain disordered dielectric superlattices. Physical Review Letters. 1987;**58**(23):2486. DOI: 10.1103/PhysRevLett.58.2486

[4] Inoue K, Ohtaka K. Photonic Crystals: Physics, Fabrication and Applications. Berlin: Springer; 2004. DOI: 10.1007/978-3-540-40032-5. ISBN 10: 6131576629

[5] Ouchani N, Bria D, Djafari-Rouhani B, Nougaoui A. Transverse-electric/transverse-magnetic polarization converter using 1D finite biaxial photonic crystal. Journal of the Optical Society of America. A. 2007; **24**(9):2710-2718. DOI: 10.1364/josaa.24.002710

[6] Mir A, Akjouj A, El Boudouti EH, Djafari-Rouhani B, Dobrzynski L. Large omnidirectional band gaps and selective transmission in one-dimensional multilayer photonic structures. Vacuum. 2001;**63**(1):197-203. DOI: 10.1016/S0042-207X(01)00191-9

[7] Ouchani N, Bria D, Djafari-Rouhani B, Nougaoui A. A birefringent reflector from a 1D anisotropic photonic crystal. Journal of Physics. Condensed Matter. 2009;**21**:485401. DOI: 10.1088/0953-8984/21/48/485401

[8] Abdulhalim I. Omnidirectional reflection from anisotropic periodic dielectric stack. Journal of Optical Communication. 2000;**174**:4350. DOI: 10.1016/S0030-4018(99)00672-0

[9] Cojocaru E. Omnidirectional reflection from multilayers with alternating birefringent and isotropic thin films. Journal of Modern Optics. 2003;**50**(9):1455-1467. DOI: 10.1080/09500340308235218

[10] Gao X, Han X, Cao W-P, Li HO, Ma HF, Cui TJ. Ultrawideband and high-efficiency linear polarization converter based on double V-shaped Metasurface. IEEE Transactions of Antennas and Propagation. 2015;**63**(8):3522-3530. DOI: 10.1109/TAP.2015.2434392

[11] Tremain B, Rance HJ, Hibbins AP, Sambles JR. Polarization conversion from a thin cavity array in the microwave regime. Scientific Reports. 2015;**5**:9366. DOI: 10.1038/srep09366

[12] Hallam BT, Lawrence CR, Hooper IR, Sambles JR. Broad-band polarization conversion from a finite periodic structure in the microwave regime. Applied Physics Letters. 2004; **84**:849851. DOI: 10.1063/1.1645661

[13] Kumar V, Singh KS, Ojha SP. One dimensional photonic band gap material as an omnidirectional reflector. Indian Journal of Physics. 2011;**85**:18111816. DOI: 10.1007/s12648-011-0204-6

[14] Chacko V, Bansal S, Hafiz AK. Effect of dispersion on omnidirectional reflection band in zinc oxide-based one-dimensional photonic crystal heterostructures. Journal of Nanophotonics. 2018;**12**(2):026012. DOI: 10.1117/1.JNP.12.026012

[15] Zhang J, Benson TM. Design of omnidirectional reflectors based on a

cascaded one-dimensional photonic crystal structure. Journal of Modern Optics. 2013;**60**:1804-1812. DOI: 10.1080/09500340.2013.863400

[16] Wang S, Yang X, Liu CT. Omnidirectional reflection in one-dimensional ternary photonic crystals and photonic heterostructures. Physics Letters A. 2014;**378**:1326-1332. DOI: 10.1016/j.physleta.2014.03.010

[17] Xiao X, Chen R. Study of omnidirectional reflection bandgap extension in one-dimensional quasi-periodic metallic photonic crystal. Journal of Nanoparticle Research. 2015; **10**(6):1550088. DOI: 10.1142/S1793292015500885

[18] Jiang L, Zheng G, Shi L, Li X. Broad omnidirectional reflector Design for Optical Telecommunications Band Using Genetic Algorithm. Japanese Journal of Applied Physics. 2008;**47**:6333. DOI: 10.1143/JJAP.47.6333

[19] Cojocaru E. Omnidirectional reflection from finite periodic and Fibonacci quasi-periodic multilayers of alternating isotropic and birefringent thin films. Journal of Applied Optics. 2002;**41**(4):747-755. DOI: 10.1364/ao.41.000747

[20] Ouchani N, Bria D, Djafari-Rouhani B, Nougaoui A. Defect modes in one-dimensional anisotropic photonic crystal. Journal of Applied Physics. 2009; **106**(11):113107-113107-8. DOI: 10.1063/1.3266005

[21] Ouchani N, El Moussaouy A, Aynaou H, El Hassouani Y, El Boudouti EH, Djafari-Rouhani B. Optical transmission properties of an anisotropic defect cavity in one-dimensional photonic crystal. Journal of Physics Letters A. 2018;**382**:231-240. DOI: 10.1016/j.physleta.2017.11.020

[22] Razi S, Ghasemi F. Broad band temperature independent photonic crystal based optical filter with response in visible wavelength range. Laser Physics. 2019;**29**:046204. DOI: 10.1088/1555-6611/ab036e

[23] Suthar B, Bhargava K, Bhargava A. Design and analysis of optical biosensor based on symmetric and asymmetric structured one-dimensional photonic crystal for tuberculosis detection. Journal of Optical Engineering. 2023;**62**(5):057104. DOI: 10.1117/1.OE.62.5.057104

[24] Zhang J, Shi J, Wang K, Gao K, Ge D, Zhang L. Effect of lithium niobate on the sensing characteristics of magnetic and thermal sensor with one-dimensional photonic crystal. Journal of Optical Engineering. 2023;**62**(7):077103. DOI: 10.1117/1.OE.62.7.077103

[25] Mousavizadeh SM, Soroosh M, Mehdizadeh F. Photonic crystal-based demultiplexers using defective resonant cavity. Optoelectronics and Advanced Materials. 2015;**9**:28-31

[26] Shokri AA, Jamshidi R. Merging of defect modes in a superlattice of one-dimensional metamaterials photonic crystals. AIP Advances. 2019;**9**:055318. DOI: 10.1063/1.5089413

[27] Wu J, Gao JX. Low temperature sensor based on one-dimensional photonic crystals with a dielectric-superconducting pair defect. Optik. 2015;**126**(24):5368-5371. DOI: 10.1016/j.ijleo.2015.09.148

[28] Mahmoodzadeh H, Rezaei B. Tunable Bragg defect mode in one-dimensional photonic crystal containing a graphene-embedded defect layer. Journal of Applied Optics. 2018;**57**: 2172-2176. DOI: 10.1364/AO.57.002172

[29] Belhadj W. Properties of omnidirectional gap and defect mode of onedimensional graphenedielectric periodic structures. Journal of Optical and Quantum Electronics. 2020;**52**:162. DOI: 10.1007/s11082-020-02267-y

[30] Goldstein H. Classical Mechanics. Addison-Wesley; 1957. p. 399. ISBN 10: 0201025108

[31] Landry GD, Maldonado TA. Complete method to determine transmission and reflection characteristics at a planar interface between arbitrarily oriented biaxial media. Journal of the Optical Society of America. A. 1995;**12**:2048-2063. DOI: 10.1364/JOSAA.12.002048

[32] Dobrzynski L. Interface response theory of continuous composite materials. Surface Science. 1987;**180**: 489-504. DOI: 10.1016/0039-6028(87) 90222-6

[33] Dobrzynski L. Interface response theory of continuous composite systems. Surface Science Reports. 1990;**11**: 139-178. DOI: 10.1016/0167-5729(90) 90003-V

[34] Ouchani N. Cristaux photoniques et phononiques anisotropes 1D. Editions Universitaires Europennes; 2011. p. 260

[35] Ouchani N, Bria D, Nougaoui A, Djafari-Rouhani B. Photonic band structure and omnidirectional band gap in anisotropic superlattice. Journal of Solar Energy Materials and Solar Cells. 2006;**90**:14451457. DOI: 10.1016/j. solmat.2005.10.011

[36] Yeh P. Optical waves in layered media. In: Wiley Series in Pure and Applied Optics. New York: Wiley; 1988. p. 20153

[37] Ouchani N, El Moussaouy A, Aynaou H, El Hassouani Y, El Boudouti EH, Djafari-Rouhani B. Superluminal and negative delay times in isotropic-anisotropic one-dimensional photonic crystal. Journal of Applied Physics. 2017;**122**:183106. DOI: 10.1063/ 1.4999095